Multiscale Characterization of Biological Systems

Vikas Tomar • Tao Qu • Devendra K. Dubey
Devendra Verma • Yang Zhang

Multiscale Characterization of Biological Systems

Spectroscopy and Modeling

 Springer

Vikas Tomar
Purdue University
West Lafayette, IN, USA

Tao Qu
Purdue University
West Lafayette, IN, USA

Devendra K. Dubey
Indian Institute of Technology Delhi
New Delhi, Delhi, India

Devendra Verma
Purdue University
West Lafayette, IN, USA

Yang Zhang
Purdue University
West Lafayette, IN, USA

ISBN 978-1-4939-3451-5   ISBN 978-1-4939-3453-9   (eBook)
DOI 10.1007/978-1-4939-3453-9

Library of Congress Control Number: 2015953098

Springer New York Heidelberg Dordrecht London

Printed on acid-free paper

Springer Science+Business Media LLC New York is part of Springer Science+Business Media (www.springer.com)

To Samarth, Yash, and Swati

Preface

Inspired by a remarkable combination of physical and mechanical properties in biological materials, a new research field has evolved that focuses on studying these materials for the underlying principles and mechanisms of operation in order to incorporate such into the engineered materials. This field or stream of materials science referred to as "Bioenabled and Biomimetic Materials" has attracted imagination of a range of scientific disciplines. The key of natural- or nature-inspired design of materials is different, organized scale levels (nano- to macroscale) of structural arrangement with the presence of the interface between the organic and inorganic phases at each level. The accurate knowledge of such design principle is needed to optimize performance of biomimetic materials for required loading condition and operation. Two aspects of this knowledge are the accurate characterization of the organic–inorganic interfaces and the quantitative study of the contributor of different interfaces and structural arrangements to the consequent improvements in mechanical properties (e.g., stiffness, strength, toughness, etc.). Another important aspect of this design is the development of an ability to better manufacture traditionally developed composites. This collection focuses on the work done by Interfacial Multiphysics Lab and collaborators on the first aspect. The field of biomimetic materials is still in its beginning and growing. This work's primary contribution is in its focus on interfaces from the point of views of experiments and multiscale models.

Authors are immensely grateful to collaborators such as Prof. Christian Hellmich at TU Vienna, Prof. Kalpana Katti at North Dakota State University, Prof. Huajian Gao at Brown University, Prof. Markus Buehler at MIT, and Prof. Glen Niebur at University of Notre Dame in indirectly or directly providing great motivation and discussions. The book contains a discussion of work by a range of researchers in this area. Permission was obtained for figures borrowed to appear in review papers and book chapters on which the book is based. We remain indebted for the related colleagues for providing required permissions.

West Lafayette, IN, USA   Vikas Tomar
West Lafayette, IN, USA   Tao Qu
New Delhi, Delhi, India   Devendra K. Dubey
West Lafayette, IN, USA   Devendra Verma
West Lafayette, IN, USA   Yang Zhang

Contents

Chapter 1
Introduction

Abstract Biological materials have evolved over millions of years. In the structural studies of biological materials, it is observed that at the mesoscale (~100 nm to few µm), the mineral crystals are preferentially aligned along the length of the organic-phase polypeptide molecules in a hierarchical (e.g., staggered or Bouligand pattern) arrangement. Multicomponent hierarchical structure of biomaterials results in the organic–inorganic interfaces involved at different length scales, i.e., between the basic components at the nanoscale, between the mineralized fibrils at the microscale, and between the layers of the multilayered structures at micro- or macroscale. Interfaces control biological reactions, provide unique organic microenvironments that can enhance specific affinities, provide self-assembly in the interface plane that can be used to orient and space molecules with precision, etc. This collection provides recent research work done in the area of interface mechanics of collagen- and chitin-based biomaterials along with various techniques that can be used to understand the mechanics of biological systems and materials.

Keywords Biological materials • Biomimetics • Computational and experimental analyses

Biological materials have evolved over millions of years. The toughness of spider silk, the strength and lightweight of bamboos, self-healing of bone, high toughness of nacre, and the adhesion abilities of the gecko's feet are a few of the many examples of high-performance biological materials. Hard biomaterials such as bone, nacre, and dentin have intrigued researchers for decades for their high stiffness, toughness, and self-healing capabilities. Such biological materials have been reviewed in appreciable detail, in the context of their hierarchical structure, material properties, and failure mechanisms [1–5]. Such hard biological materials are not only light-weight but also possess high toughness and mechanical strength. In particular, such materials combine two properties which are quite often contradictory but essential for the function of such materials. A unique feature that determines properties of such materials is interfacial interaction between organic and inorganic phases in the form of protein (e.g., chitin (CHI) or tropocollagen (TC))–mineral (e.g., calcite (CAL) or hydroxyapatite (HAP)) interfaces. The volume fraction of the protein–mineral interfaces can be enormous as the mineral bits have nanoscale size. For example, in a raindrop size volume of a nanocomposite, the area of interfacial region can be as large as a football field [6].

In the structural studies of such biological materials, it is observed that at the mesoscale (~100 nm to few μm), the mineral crystals are preferentially aligned along the length of the organic-phase polypeptide molecules in a hierarchical (e.g., staggered or Bouligand pattern) arrangement [7–11]. Interfaces are perceived to play a significant role in the stress transfer and the consequent improvements in stiffness and strength of such material systems. However, how exactly the change in interfacial chemical configuration leads to change in mechanical properties in such materials is a big subject of debate. The length scale and complexity of microstructure of hybrid interfaces in biological materials make it difficult to study them and understand the underlying mechanical principles, which are responsible for their extraordinary mechanical performance. For this reason the governing mechanisms for the mechanical behavior for such biomaterials have not been understood completely. One of the most important aspects of understanding the influence of interfaces on natural material properties is the knowledge of how stress transfer occurs across the organic–inorganic interfaces. Multicomponent hierarchical structure of biomaterials results in the organic–inorganic interfaces involved at different length scales, i.e., between the basic components at the nanoscale, between the mineralized fibrils at the microscale, and between the layers of the multilayered structures at micro- or macroscale.

From the biological/chemical perspective, interface concepts related to cell surface/synthetic biomaterial interface and extracellular matrix/biomolecule interface have wide applications in medical and biological technology. Interfaces control biological reactions, provide unique organic microenvironments that can enhance specific affinities, provide self-assembly in the interface plane that can be used to orient and space molecules with precision, etc. Interfaces also play a significant role in determining structural integrity and mechanical creep and strength properties of biomaterials. Structural arrangement of interfaces combined with interfacial interaction between organic and inorganic phases in the biomaterial interfaces significantly determines mechanical properties of biological materials, especially the aspect that leads to a unique combination of seemingly "inconsistent" properties, such as fracture strength and tensile strength both being high—as opposed to traditional engineering materials, which have high fracture strength linked to low tensile strength and vice versa.

This collection provides recent research work done in the area of interface mechanics of collagen- and chitin-based biomaterials along with various techniques that can be used to understand the mechanics of biological systems and materials. After this overview the second (next) chapter focuses on an overview of spectroscopic imaging of biological systems with special emphasis on algae as an example. Emphasis is on predicting and correlating chemistry with mechanical stress. The third chapter focuses on using a combination of electron microscopy with nanomechanics to reveal interface mechanical behavior based attributes in two different shrimp species which are chitin-based systems. The fourth chapter focuses on analyzing the role of microstructure and interfaces in the mechanical behavior based on atomistic simulations.

Insights presented in the fourth chapter are combined and presented in the form of literature phenomenological models in Chap. 5. Emphasis is on presenting existing gaps that phenomenological models cannot address. The final chapter of the present work presents an approach that could be used to correlate the continuum behavior of biological materials based on performing atomistic simulations.

References

1. P. Fratzl, R. Weinkamer, Nature's hierarchical materials. Prog. Mater. Sci. **52**, 1263–1334 (2007)
2. M.A. Meyers, P.Y. Chen, A.Y.M. Lin, Y. Seki, Biological materials: structure and mechanical properties. Prog. Mater. Sci. **53**, 1–206 (2008). doi:10.1016/j.pmatsci.2007.05.002
3. J.Y. Rho, L. Kuhn-Spearing, P. Zioupos, Mechanical properties and the hierarchical structure of bone. Med. Eng. Phys. **20**, 92–102 (1998)
4. M.E. Launey, R.O. Ritchie, On the fracture toughness of advanced materials. Adv. Mater. **21**, 2103–2110 (2009). doi:10.1002/adma.200803322
5. D.K. Dubey, V. Tomar, Role of molecular level interfacial forces in hard biomaterial mechanics: a review. Ann. Biomed. Eng. **38**, 2040–2055 (2010)
6. R. Vaia, Polymer nanocomposites: status and opportunities. MRS Bull. (USA) **26**, 394–401 (2001)
7. W.J. Landis et al., Mineralization of collagen may occur on fibril surfaces: evidence from conventional and high-voltage electron microscopy and three-dimensional imaging. J. Struct. Biol. **117**, 24–35 (1996)
8. W.J. Landis, K.J. Hodgens, J. Arena, M.J. Song, B.F. McEwen, Structural relations between collagen and mineral in bone as determined by high voltage electron microscopic tomography. Microsc. Res. Tech. **33**, 192–202 (1996)
9. P. Fratzl, N. Fratzlzelman, K. Klaushofer, G. Vogl, K. Koller, Nucleation and growth of mineral crystals in bone studied by small-angle X-ray scattering. Calcif. Tissue Int. **48**, 407–413 (1991)
10. S. Weiner, Y. Talmon, W. Traub, Electron diffraction of mollusc shell organic matrices and their relationship to the mineral phase. Int. J. Biol. Macromol. **5**, 325–328 (1983)
11. A. Al-Sawalmih et al., Microtexture and chitin/calcite orientation relationship in the mineralized exoskeleton of the American lobster. Adv. Funct. Mater. **18**, 3307–3314 (2008)

Chapter 2
Spectroscopic Experiments: A Review of Raman Spectroscopy of Biological Systems

Abstract Raman spectroscopy is fast emerging as an important characterization tool for biological systems. Raman spectroscopy has proven to be a powerful and versatile characterization tool used for determining chemical composition of material systems such as nanoscale semiconductor devices or biological systems. One major advantage of Raman spectroscopy in the case of biological molecules is that water gives very weak, uncomplicated Raman signal. Biological systems are essentially wet systems; hence, Raman spectrum of a biological system can be easily obtained by filtering the water's Raman signal. Another advantage of Raman spectroscopy in the case of biological molecules is the ability of Raman spectroscopy to analyze in vivo samples. This aspect gives this technique an edge over other methods such as infrared (IR) spectroscopy which requires elaborate signal preparation for excitation and complex instrumentation for signal processing after the excitation. This chapter focuses on presenting information on advancements made regarding the Raman spectroscopy of algae.

Keywords Raman • Biology • Algae • Material chemistry

2.1 Introduction

Raman spectroscopy has proven to be a powerful and versatile characterization tool used for determining chemical composition of material systems such as nanoscale semiconductor devices or biological systems. Raman spectroscopy is based on the concept of the Raman effect [1]. The principle behind the Raman effect is based on the inelastic scattering of incident photons by atoms and molecules in a sample. The incident photons enter a virtual energy state when they interact with sample. The eventual return of photons to ground state results in the inelastic scattering. The wavelength of scattered photons can be determined by calculating the induced dipole moments in molecules due to vibrational displacements. If the final ground state has more energy than initial state, then emitted photon will be shifted to lower frequency. This scattering is called as Stokes scattering. If the final state is more energetic than initial, the emitted photon will be shifted to higher frequency, resulting in anti-Stokes scattering. Depending on the amount of scattered photons, Raman spectrum shows

© Springer Science+Business Media New York 2015   5
V. Tomar et al., *Multiscale Characterization of Biological Systems*,
DOI 10.1007/978-1-4939-3453-9_2

various peaks which undergo changes with changes in the characteristics of a sample. These characteristic peaks can be used to identify the structural components or chemical composition of the sample.

One major advantage of Raman spectroscopy in the case of biological molecules is that water gives very weak, uncomplicated Raman signal [2]. Biological systems are essentially wet systems hence Raman spectrum of a biological system can be easily obtained by filtering the water's Raman signal. Another advantage of Raman spectroscopy in the case of biological molecules is the ability of Raman spectroscopy to analyze in vivo samples [3]. Raman spectroscopy generally does not require sample preparation for obtaining the response from varied biological samples such as algae cells [3]. This aspect gives this technique an edge over other methods such as infrared (IR) spectroscopy which requires elaborate signal preparation for excitation and complex instrumentation for signal processing after the excitation [3]. Raman spectroscopy is fast emerging as an important characterization tool for biological systems [4]. With this view, the present review focuses on presenting information on advancements made regarding the Raman spectroscopy of algae.

Algae are eukaryotic microorganisms which contain chlorophyll and are capable of photosynthesis. Algae are a rich source of carbohydrates and other nutrients [5]. Some of the important algal extracts are used in food, cosmetics, and pharmaceutical industry [5]. Algae are known to show changes in composition or the structure depending on the changes in environmental conditions [3]. Based on such attributes, algae have led to new development in applications such as sensing elements in biosensors [6–15]. Algae cells have also been used as an aid in controlling water pollution and heavy metal pollution [10, 12, 13, 15]. One of the most important and visible uses of algae has been in the domain of biofuel development due to superior lipid content in algae cells compared to other plant cells [16, 17].

Algae cells contain mainly five types of biomolecules: proteins, carbohydrates, lipids, nucleic acids, and pigments [18]. Each type of biomolecules has been shown to have its own characteristic signature Raman spectrum. Characteristic peaks of each biomolecule can be used to identify them from Raman spectrum of algae cell. Raman spectrum of an algae cell will be the sum of Raman spectra of its constituent biomolecules [16]. In various studies, Raman spectra have been used to identify a particular type of algae species in a group of different types of algae or in different environmental conditions [19–22]. Raman spectra of individual biomolecules can be used to determine the molecular structure and various properties of the biomolecules and hence have also been studied extensively. Figure 2.1a shows the overview of applications of Raman spectroscopy in study of algae. There have been several reviews of applications of Raman spectroscopy in bioanalysis [4, 23, 24]; however, none of the available review deals with Raman spectroscopy of algae cells.

This present review is divided in five parts. Introduction is provided in Sect. 2.1. In Sect. 2.2, the basic instrumentation required for performing Raman spectroscopy is presented along with recent development of advanced methods such as surface-enhanced Raman spectroscopy (SERS). In Sect. 2.3, various studies pertaining to identification of algae species using Raman spectroscopy are reviewed. In Sect. 2.4, studies of component biomolecules of algae using Raman spectroscopy are presented. Conclusion and future prospects are presented in Sect. 2.5.

a

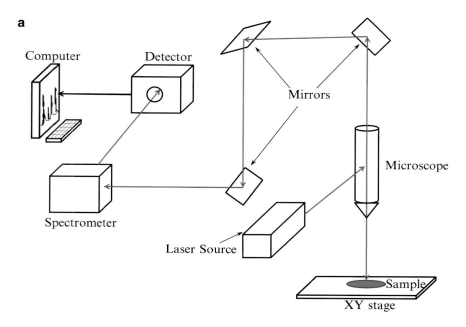

b

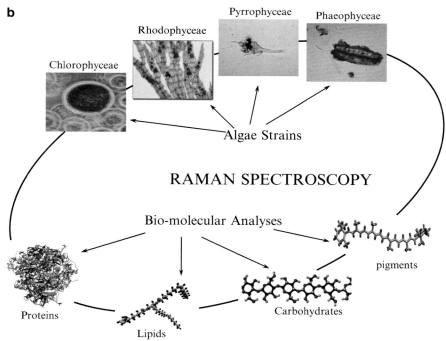

Fig. 2.1 (**a**) A schematic of experimental setup of a typical micro-Raman spectrometer and (**b**) a schematic showing the applicability of micro-Raman spectroscopy to different aspects of algae

2.2   Instrumentation

Schematic of a typical Raman setup is shown in Fig. 2.1b. In a typical Raman experiment, a laser beam excitation is provided by a laser source such as an argon ion laser or Nd:YAG laser. Wavelengths of laser lines of argon ion laser and Nd:YAG lasers are most common incident wavelengths used to study algae Raman response. These wavelengths include 488, 514.5 and 1024 nm among other wavelengths. The laser signal is focused on the sample that is to be analyzed using optical systems such as a microscope and lens–mirror assembly, and scattered response from the sample is recorded. The scattered response has wavelength higher or lower than that of the original laser depending on whether the scattered response follows Stokes or anti-Stokes scattering. The scattered beam is directed through a series of filters to obtain Raman response in the form of a spectrum of lines with varying intensity as a function of wavelength. This signal is then recorded on a computer for further processing and analyses [25].

Several advances in instrumentation have resulted in the more applications of Raman spectroscopy in biological samples [4]. Better lasers, filters, and fiber optics have improved Raman signals because of improvements in excitation of sample and better acquisition of response signals [4]. Incorporation of diode laser as optical pump in Nd:YAG lasers has provided better pump stability and hence reduced flicker noise, which increases overall signal-to-noise ratio [4]. Ti/sapphire and other solid-state lasers are also used as excitation sources for Raman spectroscopy because of their high harmonic conversion efficiencies and ability to tune them to different wavelengths [4]. Along with advances in laser systems, tunable filters such as liquid crystal and acousto-optic filters have introduced electronically controlled, fixed optical element which produces 2D images without surface or beam rastering [4]. Liquid crystals have high throughput, large spectral range, and fast scan times, making them ideal for use in the spectroscopy of biological samples. Incorporation of fiber optics for delivering the optical excitation and collection of response signal has decreased the size and increased the flexibility of the Raman systems [4]. Improved hardware gives better spatial resolution and portability to the Raman system, hence making in vivo analyses easier.

Several new discoveries in the physics of Raman effect have resulted in the advances in the Raman spectroscopic technique for bioanalysis. Resonance Raman spectroscopy uses the lasers which have their energy adjusted such that the energy of the laser or that of the scattered signal coincides with electronic transition energy of a particular molecule or crystal [26]. This technique has become more prevalent with the advances in tunable lasers [4]. In biological samples, the advantage of resonance Raman spectroscopy is that the only modes associated with specific chromophoric group of molecules are enhanced due to resonance effect [26]. Resonance Raman spectroscopy makes it possible to study particular molecule in the algae cell, thus enabling analysis of individual component from cell. Recently, there have been some major advances in Raman spectroscopy techniques to analyze the biological samples. These include surface-enhanced Raman spectroscopy (SERS), tip-enhanced Raman spectroscopy (TERS), coherent anti-Stokes Raman spectroscopy (CARS), and laser tweezers Raman spectroscopy (LTRS).

SERS has become an important Raman spectroscopic analysis tool because it offers 10^3–10^7 enhancement in the intensity of the Raman response signal of an analyte when the analyte is adsorbed on the surface of some noble metals with nanoscale features [23]. Surface enhancement permits a single molecule detection as large intensity enhancement in response signal makes the response signal detectable even when response is obtained from a single molecule [23] and improves the spatial resolution to lateral resolutions better than 10 nm [27]. Information about the surface–interface processes can also be obtained using the extent of increase in intensity of each Raman mode [27]. The enhancement in intensity of the response is not observed for any other metal colloids or surfaces having large features [28]. The enhancement in response signal intensity is because of two different mechanisms. First, increase in the intensity due to electromagnetic enhancement correlated with the excitation of surface plasmons in metal structures and second, intensity increase due to chemical enhancement related to the adsorbate–substrate complexes providing different orbitals for excitation of the Raman processes [28]. Surface-enhanced resonance Raman spectroscopy (SERRS) combines the advantages offered by resonance Raman spectroscopy and surface enhancement. SERRS can be used for higher specificity in excitation, thus opening interesting opportunities for studying specific parts of large biomolecules [27]. SERS and SERRS make it possible to analyze single cell of the algae. It is also possible to study individual molecules in algae cells using SERS.

The concept of TERS is based on the concept of "hot spots" in surface-enhanced Raman spectroscopy [28]. This concept states that the surface enhancement is produced by small Raman active areas called "hot spots" on the surface and the rest of the surface is inactive [28]. In TERS, such single hot spot is created by keeping a sharp protrusion or a tip made of gold or silver at a small distance from the sample [28]. The enhancement in intensity in TERS is because of localized enhanced electromagnetic field near the tip apex [28]. The enhanced electromagnetic field provides the enhancement in the intensity of Raman spectrum like in SERS. Currently, TERS provides two to four orders of magnitude increase in response signal intensity. The response intensity enhancement is expected to increase further on improvements in excitation and radiation efficiency in surface plasmons [28]. TERS provides better spatial resolution compared to traditional Raman spectroscopy due to local enhancement provided by hot spots, thus making the single-cell analysis possible.

In CARS, two high-powered (generally pulsed) laser beams are focused together on the sample. Due to the mixing of two lasers, a coherent beam resembling low-intensity laser beam is generated [29]. Due to the coherence of the beam, the spectrum obtained using CARS is orders of magnitude stronger than Raman spectrum obtained using traditional Raman spectroscopy [29]. Two interacting beams give high 3D sectioning capability [23]. CARS signals can also be easily picked in fluorescent background due to blue shift of the response signals [23].

LTRS utilizes laser tweezers to immobilize a single cell of algae or any other organism [17]. LTRS can be used to obtain the Raman spectrum of single cell or even a particular part of an immobilized algae cell [17].

Advances in the spectroscopic techniques have opened a new frontier of very high-resolution Raman spectroscopy, even in the specific parts inside the cell. They also enable the monitoring of real-time changes in algae cells using Raman spectroscopy, thus improving the ability to detect the changes in algae cells in different environments. In the next section, a summary of the studies done on the identification of algae species using Raman spectroscopy is provided.

2.3   Identification of Algae Species Using Raman Spectroscopy

Several studies focusing on the identification of algae using Raman spectroscopy have been reported in literature [19–22]. The technique of identification of algae uses the characteristic peaks in the Raman spectra to identify unique biomolecules in the cells which are correlated to the specific species of algae [19, 20].

Identification process often involves analysis of Raman spectra from multiple components of cells. Analysis of multicomponent systems is performed using statistical multivariate analysis. Most of the times, linear analysis is used [30]. Two basic assumptions are made in linear analysis. First, Raman spectrum of mixture of biomolecules is assumed to be linear superposition of component spectra of biomolecules in mixture [30]. It is also assumed that signal intensity and concentration of biomolecule in mixture have linear relationship [30]. For finding major components, either explicit or implicit methods are applied. In explicit methods, such as ordinary least squares or classical least squares, Raman spectra of all component biomolecules are previously known [31]. Most of the times, Raman spectra for components are not known beforehand, and hence, implicit analysis is performed to analyze the components. Principal component analysis [32] and partial least squares [31] method are some of the most common implicit analysis methods. Statistical analyses help in extracting the information from Raman spectrum and analyze the component mixtures effectively.

In Ref. [22], Raman spectroscopy is applied to differentiate between nontoxic and toxic algal strains. Four different species of algae including *Pseudo-nitzschia*, some of which are capable of producing toxin domoic acid, are studied by the means of resonance Raman spectra excited at 457.9 and 488 nm. It was observed that Raman spectra for all algae species contain major peaks near 1000, 1153, and 1523 cm^{-1}, all of which are strongly enhanced due to carotenoids [22]. Features between 920–980 and 1170–1230 cm^{-1} are relatively weaker and are more characteristic of algae species [22]. It was observed that all *Pseudo-nitzschia* species produce a Raman response signal which is different from other species [22]. In the study by Brahma et al. [19], various marine algae species were identified from the medium containing different algae species using resonance Raman spectroscopy. In Ref. [20], different types of seaweeds were identified using Fourier transform-based Raman spectroscopy (FT-Raman). It was observed that FT-Raman spectra have higher resolution than FTIR spectra; hence, it was concluded that FT-Raman spectroscopy is better for identification of different species from a medium containing different species of algae. In Ref. [21], algae and bacteria deposits on Ti sheets are detected using surface-enhanced Raman spectroscopy. Spatial distribution of algae on the surface was also identified.

Raman spectroscopy provides a high-resolution method with high accuracy for identification and detection of algae species. It can be used to detect the algae species from the biomass coatings or in algal blooms. With smaller and portable equipments, it is possible to detect the algae species in their natural habitat [33]. Recent developments in instrumentation and spectroscopic techniques such as SERS will further help in identification of algae. The identification of algae using Raman spectroscopy is based on the identification of component biomolecules. Hence, it is important to study the individual biomolecules using Raman spectroscopy. These studies are summarized in the next section.

2.4   Study of Component Biomolecules

Algae cell contains mainly five types of biomolecules: proteins, carbohydrates, lipids, nucleic acids, and pigments [18]. Raman spectroscopy has been used to find the molecular structure of the biomolecules [34–43] and to analyze other properties such as the location of biomolecules in the cell [16, 44–52]. Various methods are used to predict the molecular structure of biomolecules using Raman spectroscopy. In isotope labeling, certain atoms in molecule are replaced by their isotopes, and changes in the Raman spectra are observed for the new isotope-labeled molecules. The bands which show differences from original spectra are assigned to the labeled parts of the molecule. In site-directed mutagenesis, specific sites of cells are modified using mutation. The changes in Raman spectra can then be assigned to the mutated parts of cells and hence the biomolecules that are present in mutated parts or to the changes in biomolecules due to mutations. In normal coordinate analysis, the vibrational modes are analyzed theoretically using quantum mechanics of coupled harmonic oscillators [53], and each mode is assigned to a peak in Raman spectrum. Now, because of advances in high-performance computing, it is possible to find the vibrational modes associated with each peak in Raman spectrum of each biomolecule using molecular simulations such as density functional theory.

The reference Raman spectra for important biomolecules are collected in [54]. The reviews on the use of Raman spectroscopy in the analysis of biomolecules are presented in various references [4, 23, 24]. The summary of Raman spectroscopic studies of each type of biomolecule in algae is presented in Table 2.1.

Table 2.1 Studies on biomolecules in algae using Raman spectroscopy

Biomolecule	Algae species studied	Observations	Conclusions	Ref.
Proteins				
Adhesion proteins	*Coccomyxa* sp. and *Glaphyrella trebouziodes*	Amide III band peaks at 1224 and 1260 cm^{-1}	Consistent with generic amyloid structure with strong hydrophobic core	[55]
Hemoglobin	*C. eugametos*	Fe-CN ferric derivative peaks at 440 cm^{-1}	Fe-CN moiety adopts highly bent structure due to H bonding	[38]
		1502 and 1374 cm^{-1} peaks	Six-coordinated low-spin heme iron with distal ligand tyrosinate	
Carbohydrates				
Polysaccharides		Peaks in the range 350–600 cm^{-1}	Skeletal pyranose ring modes Glycosidic stretching modes CH2 and C-OH deformations	[44, 56]
β-D-Glucosides		377 cm^{-1} characteristic peak		[44, 56]
α-D-Glucosides		Strong peaks from 479 to 483 cm^{-1} Peak at 543 cm^{-1}	Peaks from amylopectin and amylase Dextran peak	[44, 56]
Alginates		Peaks at <1300 cm^{-1} Peaks at >1300 cm^{-1} (Fig. 2.2)	Vibration of polymer backbone Stretching vibrations of carboxylate groups	[46]
Calcium alginate		Change of band position from alginates Most prominent: 1413–1433 cm^{-1}	Symmetric COO$^-$ stretching peak	[46]
κ-Carrageenan and ι-carrageenan	*Eucheuma cottonii* *Eucheuma spinosum*	Various peaks from 700 to 1200 cm^{-1}	Peaks are assigned to respective molecular structure of carrageenans	[48]
Lipids				
Hydrocarbons (Fig. 2.3)	*Botryococcus braunii*	1650–1670 and 2800–3000 cm^{-1} peaks	Double-bond stretching peak in long-chain unsaturated hydrocarbons	[57]
	Chlorella sorokiniana *Neochloris oleoabundans*	1650 cm^{-1} peak 2800–3000 cm^{-1} peak	C=C stretching peak C=C-H vibration peak	[16]

Component	Organism	Peaks	Interpretation	Reference
	Botryococcus sudeticus, Chlamydomonas sp., Trachydiscus minutus	1656 cm⁻¹ peak 1445 cm⁻¹ peak	Cis C=C stretching mode proportional to no. of unsaturated bonds CH₂ scissoring mode proportional to no. of saturated bonds Trachydiscus minutus has significantly higher content of unsaturated fatty acids	[56]
Pigments: chlorophylls				
Chl a (Fig. 2.4)		Peaks in the range 1100–1600 cm⁻¹ Peaks in the range 700–950 cm⁻¹ Peaks <700 cm⁻¹ and from 900 to 1000 cm⁻¹ Strong 306 cm⁻¹ peak and absence of 317 cm⁻¹ peak	Stretching motions of C–C and C–N bonds Planar deformations of tetrapyrrole macrocycle Mg-related vibrations Mg atoms in hexacoordinated state	[35]
Chl b		Peaks similar to Chl a	Chl a and Chl b have very similar structure	[35]
Chl c (Fig. 2.5)		1361 cm⁻¹ peak	C–N ring breathing mode	[58]
Chl d (Fig. 2.6)	Acaryochloris marina	Very different Raman response as compared to Chl a and Chl b	Presence of formyl group at C-3 position	[37]
Pigments: carotenes				
β-Carotene (Fig. 2.7)	Chlorella sorokiniana, Neochloris oleoabundans, Euglena, Chlamydomonas	Peak at 1520 cm⁻¹ Peak at 1157 cm⁻¹ Peak at 1000 cm⁻¹ Strong peak in the eyespot region	C=C stretching mode C–C stretching mode C–CH3 stretching mode High content of carotenoids in eyespot region	[16, 51, 52, 59, 60]
Nucleic acids				
DNA and RNA		Peaks between 600 and 800 cm⁻¹ Intense peak at 1671 cm⁻¹ Peak at 1100 cm⁻¹	Ring breathing modes of DNA and RNA bases C=O stretch vibrations Symmetric PO₂⁻ stretching vibration	[54]

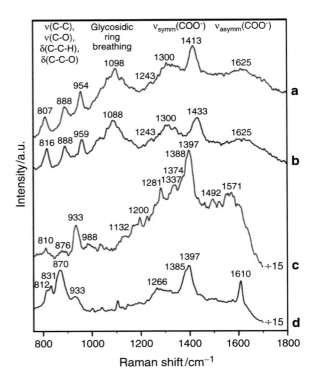

Fig. 2.2 (a) Raman spectrum of Na alginate. (b) Raman spectrum of Ca alginate. (c) Raman spectrum of Na alginate mixed with silver colloid. (d) Raman spectrum of Ca alginate mixed with silver colloid [46]

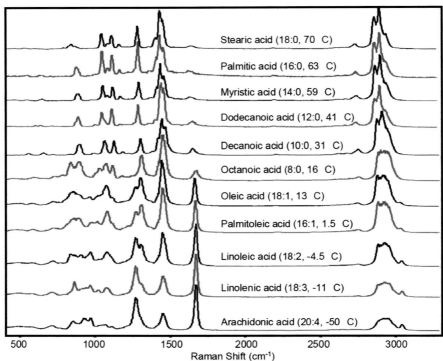

Fig. 2.3 Raman spectra of various lipid molecules [17]

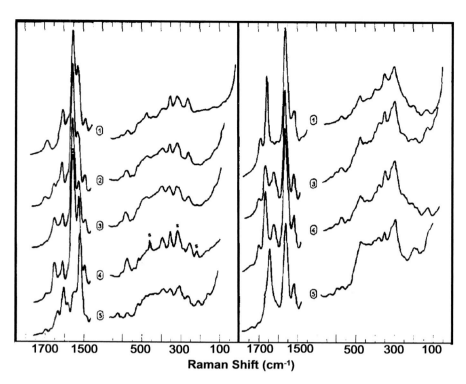

Fig. 2.4 Raman spectra for Chl a (*left*) and Chl b (*right*) in different solvents. (**1**) Acetone. (**2**) Hexane/acetone, 90: 1Ov/v. (**3**) Chl a: dry hexane; Chl *b*: dry cyclohexane. (**4**) Dry carbon tetrachloride. (**5**) Water aggregates in Nujol oil [35]

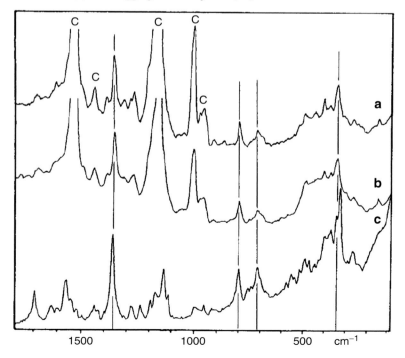

Fig. 2.5 (**a**) Raman spectra for whole algae cells of algae *Phaeodactylum tricornutum* (**b**) and *Gomphonema parvulum* and (**c**) chlorophyll c molecule [36]

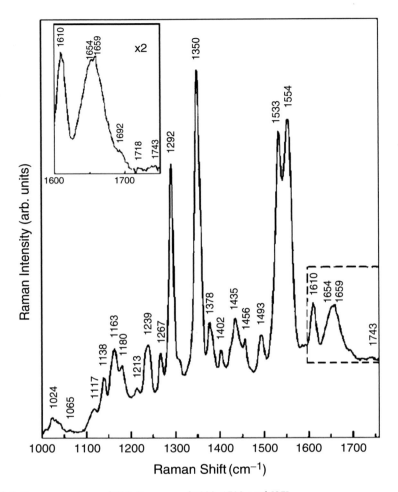

Fig. 2.6   Raman spectrum of Chl *d* in range of 1000–1700 cm^{-1} [37]

2.5   Conclusions

Raman spectroscopy is an important tool in the analyses of algae cells and component biomolecules. It is an ideal experimental measurement system for the characterization of "wet" biosystems on the account of weak Raman signal of water. It also provides high spatial and temporal resolution required for the characterization of biological samples at cellular level. Raman spectrum also has a lot of information focusing on vibration states of the molecules. Any single Raman spectrum can be used to extract large amount of data about the systems including components of system and temperature of system. Using the information contained in the spectrum,

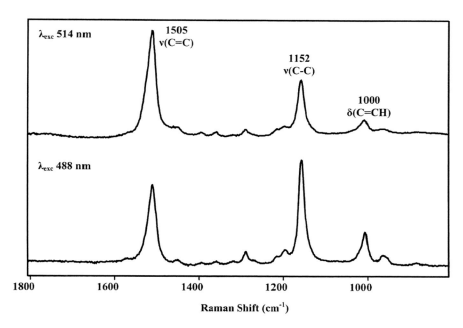

Fig. 2.7 Raman spectra of bacteria showing the carotenoid peaks at 1505, 1152, and 1000 cm⁻¹ [60]

various algae species can be identified and analyzed for various properties such as locations of particular biomolecule in the cell.

With advances in spectroscopic techniques in the form of SERS, TERS, CARS, and LTRS, it is now possible to obtain Raman spectra of specific parts of a cell. This will enable better understanding of particular parts of cells and hence the constituent biomolecules. The use of pulsed lasers has enabled monitoring of the changes in system with respect to time. In the future, Raman spectroscopy can be used to map the real-time changes in the biological systems with very high spatial resolution and also very high accuracy.

Raman spectroscopy can expand its application base in the future including clinical diagnosis because of its various advantages. The most important applications of Raman spectroscopy in the study of algae in the future could well be for studying in vivo samples and for the analysis of algae cell metabolism in real time.

References

1. C.V. Raman, K.S. Krishnan, A new type of secondary radiation. Nature **121**, 501–502 (1928)
2. S. Frank, E.R. Parker, *Applications of Infrared, Raman, and Resonance Raman Spectroscopy in Biochemistry* (Springer, New York, 1983)
3. P. Heraud, B.R. Wood, J. Beardall, D. McNaughton, *New Approaches in Biomedical Spectroscopy* (American Chemical Society, Washington, DC, 2007)

4. D. Pappas, B.W. Smith, J.D. Winefordner, Raman spectroscopy in bioanalysis. Talanta **51**, 131–144 (2000)
5. K.H.M. Cardozo et al., Metabolites from algae with economical impact. Comp. Biochem. Physiol. C **146**, 60–78 (2007)
6. B. Podola, E.C.M. Nowack, M. Melkonian, The use of multiple-strain algal sensor chips for the detection and identification of volatile organic compounds. Biosens. Bioelectron. **19**, 1253–1260 (2004)
7. P. Mayer, R. Cuhel, N. Nyholm, A simple *in vitro* fluorescence method for biomass measurements in algal growth inhibition tests. Water Res. **31**, 2525–2531 (1997)
8. R.G. Smith, N. D'Souza, S. Nicklin, A review of biosensors and biologically-inspired systems for explosives detection. Analyst **133**, 571–584 (2008)
9. M. Campas, R. Carpentier, R. Rouillon, Plant tissue — and photosynthesis — based biosensors. Biotechnol. Adv. **26**, 370–378 (2008)
10. C. Durrieu, C. Tran-Minh, Optical algal biosensor using alkaline phosphatase for determination of heavy metals. Ecotoxicol. Environ. Saf. **51**, 206–209 (2002)
11. M. Naessens, J.C. Leclerc, C. Tran-Minh, Fiber optic biosensor using *chlorella vulgaris* for determination of toxic compounds. Ecotoxicol. Environ. Saf. **46**, 181–185 (2000)
12. D. Franse, A. Muller, D. Beckmann, Detection of environmental pollutants using optical biosensor with immobilized algae cells. Sensors Actuators B **51**, 256–260 (1998)
13. A. Scardino et al., Delayed luminescence of microalgae as an indicator of metal toxicity. J. Phys. D. Appl. Phys. **41**(155507), 155501–155508 (2008)
14. Y. Tanaka et al., Biological cells on microchips: new technologies and applications. Biosens. Bioelectron. **23**, 449–458 (2007)
15. L. Campanella, F. Cubadda, M.P. Sammartino, A. Saoncella, An algal biosensor for the monitoring of water toxicity in estuarine environments. Water Res. **35**, 69–76 (2000)
16. Y.Y. Huang, C.M. Beal, W.W. Cai, R.S. Ruoff, E.M. Terentjev, Micro-Raman spectroscopy of algae: composition analysis and fluorescence background behavior. Biotechnol. Bioeng. **105**, 889–898 (2010)
17. H. Wu et al., *In-vivo* lipidomics using single-cell Raman spectroscopy. Proc. Natl. Acad. Sci. U. S. A. **108**, 3809–3814 (2011)
18. B. Alberts et al., *Essential Cell Biology* (Garland Publishing Inc., New York, 1998)
19. S.K. Brahma, P.E. Hargraves, W.F. Howard, W.H. Nelson, A resonance Raman method for the rapid detection and identification of algae in water. Appl. Spectrosc. **37**, 55–58 (1983)
20. L. Pereira, A. Sousa, H. Coelho, A.M. Amado, P.J.A. Ribeiro-Claro, Use of FTIR, FT-Raman and C-NMR spectroscopy for identification of some seaweed phycocolloids. Biomol. Eng. **20**, 223–228 (2003)
21. S. Ramya, R.P. George, R.V. Subba Rao, R.K. Dayal, Detection of algae and bacterial biofilms formed on titanium surfaces using micro-Raman analysis. Appl. Surf. Sci. **256**, 5108–5105 (2010)
22. Q. Wu et al., Differentiation of algae clones on the basis of resonance Raman spectra excited by visible light. Anal. Chem. **70**, 1782–1787 (1998)
23. J.R. Baena, B. Lendl, Raman spectroscopy in chemical bioanalysis. Curr. Opin. Chem. Biol. **8**, 534–539 (2004)
24. H. Fabian, P. Abzenbacher, New developments in Raman spectroscopy of biological systems. Vib. Spectrosc. **4**, 125–148 (1993)
25. I. De Wolf, Micro-Raman spectroscopy to study local mechanical stress in silicon integrated circuits. Semicond. Sci. Technol. **11**, 139–154 (1996)
26. D.P. Strommen, K. Nakamoto, Resonance Raman spectroscopy. J. Chem. Educ. **54**, 474–478 (1977)
27. K. Kneipp, H. Kneipp, I. Itzkan, R.R. Dasari, M.S. Feld, Surface-enhanced Raman scattering and biophysics. J. Phys. Condens. Matter **14**(R), 597–624 (2002)
28. B. Pettinger, in *Topics in Applied Physics: Surface-Enhanced Raman Scattering-Physics and Applications*, ed. by C. Ascheron, vol. 103 (Springer, New York, 2006), pp. 217–240
29. W.M. Tolles, J.W. Nibler, J.R. McDonald, A.B. Harvey, A review of the theory and application of Coherent Anti-Stokes Raman Spectroscopy (CARS). Appl. Spectrosc. **31**, 253–271 (1977)

30. E.B. Hanlon et al., Prospects for in vivo Raman spectroscopy. Phys. Med. Biol. **45**, R1–R59 (2000)
31. D. Haaland, M. & Thomas, E. V. Partial least-squares methods for spectral analyses. 1. Relation to other quantitative calibration methods and the extraction of qualitative information. Anal. Chem. **60**, 1193–1202 (1988)
32. S. Wold, K. Esbensen, P. Geladi, Principal component analysis. Chemom. Intell. Lab. Syst. **2**, 37–52 (1987)
33. B.R. Wood et al., A portable Raman acoustic levitation spectroscopic system for the identification and environmental monitoring of algal cells. Anal. Chem. **77**, 4955–4961 (2005)
34. M. Lutz, Antenna chlorophyll in photosynthetic membranes: a study by resonance Raman spectroscopy. Biochim. Biophys. Acta **460**, 408–430 (1977)
35. M. Lutz, Resonance Raman spectra of chlorophyll in solution. J. Raman Spectrosc. **2**, 497–516 (1974)
36. W.D. Wagner, W. Waidelich, Selective observation of chlorophyll c in whole cells of diatoms by resonant Raman spectroscopy. Appl. Spectrosc. **40**, 191–196 (1986)
37. M. Chen, H. Zeng, A.W.D. Larkum, Z.-L. Cai, Raman properties of chlorophyll d, the major pigment of *Acaryochloris marina*: studies using both Raman spectroscopy and density functional theory. Spectochim. Acta A **60**, 527–534 (2004)
38. T. Egawa, S.-R. Yeh, Structural and functional properties of hemoglobins from unicellular organisms as revealed by resonance Raman spectroscopy. J. Inorg. Biochem. **99**, 72–96 (2005)
39. P.D. Vasko, J. Blackwell, J.L. Koenig, Infrared and Raman spectroscopy of carbohydrates. Part I: Identification of O-H and C-H related vibrational modes for D-glucose, maltose, cellobiose and dextran by deuterium-substituted methods. Carbohydr. Res. **19**, 297–310 (1971)
40. P.D. Vasko, J. Blackwell, J.L. Koenig, Infrared and Raman spectroscopy of carbohydrates. Part II: Normal coordinate analysis of α-D-glucose. Carbohydr. Res. **23**, 407–416 (1972)
41. J.J. Cael, J.L. Koenig, J. Blackwell, Infrared and Raman spectroscopy of carbohydrates. Part III: Raman spectra of the polymorphic forms of amylose. Carbohydr. Res. **29**, 123–134 (1973)
42. J.J. Cael, J.L. Koenig, J. Blackwell, Infrared and Raman spectroscopy of carbohydrates. Part IV: Normal coordinate analysis of V-Amylose. Biopolymers **14**, 1885–1903 (1975)
43. J.J. Cael, K.H. Gardner, J.L. Koenig, J. Blackwell, Infrared and Raman spectroscopy of carbohydrates. Paper V: Normal coordinate analysis of cellulose I. J. Chem. Phys. **62**, 1145–1153 (1975)
44. L. Yang, L.-M. Zhang, Chemical structural and chain conformational characterization of some bioactive polysaccharides isolated from natural sources. Carbohydr. Polym. **76**, 349–361 (2009)
45. R.H. Atalla, J.M. Hackney, in *Material Research Society Symposium* (Materials Research Society, Pittsburgh, 1992)
46. T. Schmid, A. Messmer, B.-S. Yeo, W. Zhang, R. Zenobi, Towards chemical analysis of nanostructures in biofilms II: tip-enhanced Raman spectroscopy of alginates. Anal. Bioanal. Chem. **391**, 1907–1916 (2008)
47. P.D.A. Pudney, T.M. Hancewicz, D.G. Cunnigham, M.C. Brown, Quantifying the microstructure of soft solid materials by confocal Raman spectroscopy. Vib. Spectrosc. **34**, 123–135 (2004)
48. T. Malfait, H. Van Dael, F. Van Cauwelaert, Molecular structure of carrageenans and kappa oligomers: a Raman spectroscopic study. Int. J. Biol. Macromol. **11**, 259–264 (1989)
49. S. Yu et al., Physico-chemical characterization of floridean starch of red algae. Starch **54**, 66–74 (2002)
50. A. Enejder, C. Brackmann, F. Svedberg, Coherent Anti-Stokes Raman scattering microscopy of cellular lipid storage. IEEE J. Sel. Top. Quantum Electron. **16**, 506–515 (2010)
51. B. Robert, in *The Photochemistry of Carotenoids*, ed. by H.A. Frank, A.J. Young, G. Britton, R.J. Cogdell (Kluwer Academic Publishers, Dordrecht, 1999), pp. 189–201. Chapter 10
52. Y. Kubo, T. Ikeda, S.-Y. Yang, M. Tsuboi, Orientation of carotenoid molecules in the eyespot of alga: *in situ* polarized resonance Raman spectroscopy. Appl. Spectrosc. **54**, 1114–1119 (2000)

53. P. Groner, in *Handbook of Vibrational Spectroscopy* (Wiley and Sons, Chichester, UK, 2006)
54. J. De Gelder, K. De Gussem, P. Vandenabeele, L. Moens, Reference database of Raman spectra of biological molecules. J. Raman Spectrosc. **38**, 1133–1147 (2007)
55. A.S. Mostaert et al., Characterisation of amyloid nanostructures in the natural adhesive of unicellular subaerial algae. J. Adhes. **85**, 465–483 (2009)
56. O. Samek et al., Raman microspectroscopy of individual algal cells: sensing unsaturation of storage lipids *in vivo*. Sensors **10**, 8635–8651 (2010)
57. C. Largeau, E. Casadevall, C. Berkaloff, P. Dhamelincourt, Sites of accumulation and composition of hydrocarbons in *Botryococcus braunii*. Phytochemistry **19**, 1043–1051 (1980)
58. M. Wagner, Single-cell ecophysiology of microbes as revealed by Raman microspectroscopy or secondary ion mass spectroscopy imaging. Annu. Rev. Microbiol. **63**, 411–432 (2009)
59. J.C. Merlin, Resonance Raman spectroscopy of carotenoids and carotenoid-containing systems. Pure Appl. Chem. **57**, 785–792 (1985)
60. C.P. Marshall et al., Carotenoid analysis of halophilic archaea by resonance Raman spectroscopy. Astrobiology **7**, 631–643 (2007)

Chapter 3
Nanomechanics Experiments: A Microscopic Study of Mechanical Property Scale Dependence and Microstructure of Crustacean Thin Films as Biomimetic Materials

Abstract Most recent studies on the natural material include shrimp exoskeleton, crab exoskeletons, lobsters, ganoid scale of an ancient fish, toucan beak, and seashells such as nacre and mollusk. Studies focusing on biomimetic materials include development of biomimetic scaffolds for tissue growth and fabrication of tissues from biocompatible, biodegradable polymers, development of the honeycomb plates with design from beetle forewings to eliminate problems of edge sealing, molding process by thoroughly investigating beetle forewing to be able to mimic its design for better sandwich panel structures, and development of high-performance functional nanocomposites from graphene sheets with enhanced thermal conductivity and mechanical stiffness. In the present chapter, basic design principles of the crustaceans and deformation mechanisms responsible for higher strength, stiffness, and toughness are highlighted.

Keywords Nanoindentation • Nanomechanics • Microscopy • Small-scale biological mechanics • Crustacean • High-temperature shrimp

Natural materials that have developed through millions of years have a hierarchical assembly of its constituent materials. Such assembly provides them exceptional strength, toughness, and stiffness compared to their counterparts [1–9]. Most recent studies on the natural material include shrimp exoskeleton [3, 4], crab exoskeletons [10–13], lobsters [1, 2, 14, 15], ganoid scale of an ancient fish [16], toucan beak [17, 18], and seashells such as nacre and mollusk [19–28]. These studies have revealed interesting features in the design of such biocomposites that make them much stronger than the constitutive materials. The exoskeleton of crustaceans such as shrimps, lobsters, and crabs is in the form of multiple thin films stacked in a Bouligand pattern with a well-defined hierarchical structure [1, 10, 12–15, 26, 29]. It consists of chitin-based fibrils coated in proteins at nanometer level. Such fibrils bind together to form fibers. These fibers are then woven together to form chitin–protein thin-film layers. These layers are stacked in a twisted plywood structure known as the Bouligand pattern [1, 10, 12–15, 26, 29]. The spacing between such woven layers is filled with proteins and biominerals. Studies by Seki et al. [17, 18] have concentrated on a similar kind of exoskeleton found in the beak of toucan. It was found to be a sandwich structure with the exterior of keratin and a closed cell fibrous network made of calcium-rich proteins. Raabe et al. [1, 14] have studied structure and mechanical properties of lobster and crab exoskeletons. They reported hardness and reduced stiffness changing with depth of examination in exoskeleton of lobster. The exoskeleton features a graded design with different stacking densities of chitin-based Bouligand structure. Boßelmann et al. [15] have shown a direct correlation between increase in mineral content and hardness of lobster claw. Chen et al. [10] compared the mechanical properties of the crab shell in dry and wet conditions. Exocuticle was found to be more dense and harder than endocuticle, with the presence of water leading to increased toughness. Melnick et al. [30] have investigated the effect of dark pigment of stone crab claws on its mechanical properties. Lian et al. [11] reported mechanical properties of dungeon crab exoskeleton. Our previous experimental work on shrimp exoskeletons has been focused on understanding the temperature effect [3, 4]. Studies focusing on biomimetic materials include development of biomimetic scaffolds for tissue growth and fabrication of tissues from biocompatible, biodegradable polymers [31, 32], development of the honeycomb plates with design from beetle forewings to eliminate problems of edge sealing, molding process by thoroughly investigating beetle forewing to be able to mimic its design for better sandwich panel structures [33–36], and development of high-performance functional nanocomposites from graphene sheets with enhanced thermal conductivity and mechanical stiffness [37, 38].

A lot of mechanistic studies have been performed in order to study the influence of biological microstructure on biomaterial strength. Feng et al. found that the crack deflection, fiber pullout, and organic matrix bridging are the three main toughening mechanisms acting on nacre. Along with it was found that the organic matrix also plays an important role in the toughening of this biological composite. This principle of structure and property to the synthesis of structural ceramic materials can be applied to improve the toughness of ceramics [25]. Chen et al. investigated the structure of natural ceramic mollusk shell for fracture strength and fracture toughness. It has different shapes and arrangements of laminated aragonites and organic layers. The arrangements of aragonites have also various forms which include parallel, crossed, and inclined ones. The size, shape, and arrangement adopted depend strongly on the state of local stress [22]. Gao et al. showed that the aspect ratio and the staggered alignment of mineral crystals are the key factors contributing to the high stiffness of biocomposites in spite of the extremely soft protein constituent. The mineral aspect ratio should have an optimum value to balance the stiffness, the strength of mineral crystals, the strength of protein, the strength of interface, the fracture energy, and the viscoelastic properties of biocomposites [21]. Schneider et al. developed models that are capable of predicting strength values for real biomaterials up to five hierarchical levels by extracting information about the mechanical properties at different hierarchical levels from these experimental data [7]. Mayer studied the mechanisms underlying the toughening in rigid natural composites exhibited by the concentric cylindrical composites of spicules of sponges and by the nacre (brick-and-mortar) structure of mollusks [8]. Kumar et al. performed structural characterization studies to understand the construction of the exoskeleton of barnacles [28].

In the present paper, basic design principles of the crustaceans and deformation mechanisms responsible for higher strength, stiffness, and toughness are highlighted at different hierarchical levels based on information obtained using imaging techniques such SEM (scanning electron microscopy), EDX (energy-dispersive X-ray), and mechanical characterization techniques such as nanoindentation and AFM (atomic force microscopy) using example of two shrimp species *Rimicaris exoculata* and *Pandalus platyceros*.

3.1   Methods

Nanoindentation is one of the preferred methods for characterization of mechanical properties of materials at micro-nanoscale. The material being investigated (shrimp exoskeleton) is very thin and heterogeneous. In such material, the traditional uni-axial mechanical loading tests will only provide overall mechanical information, while nanoindentation has the capability to give site-specific data with minimal sample preparation. This makes nanoindentation a convenient experimental technique to measure elastic modulus and hardness at nano- and microscale.

3.1.1   Nanoindentation

Figure 3.1 shows a standard indentation curve and parts of curve used for material property calculations. The experimental procedure involves indenting the surface of material being tested by increasing force in small steps until peak load (P_{max}) or peak depth (h_{max}) is achieved. The Berkovich indenter was used in the present work. The unloading part was used for predicting the material properties using a framework based on contact mechanics [39, 40]. During experiments, maximum

Fig. 3.1 Indentation curve showing parts of curve used for stiffness, creep, and thermal drift calculations

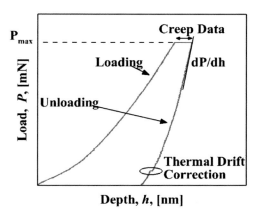

indentation load P_{max} and maximum area of indentation A are measured. The hardness H is given by

$$H = \frac{P_{max}}{A}.$$ (3.1)

The parameter A for an ideal Berkovich indenter used in the present work as a function of contact depth h_c is given as

$$A = 3\sqrt{3}h_c^2 \tan^2 65.3° \approx 24.5h_c^2.$$ (3.2)

The reduced Young's modulus E_r is related to the slope of the upper part of the unloading curve by

$$S = \frac{dP}{dh} = 1.17E_r\sqrt{A}.$$ (3.3)

Here, S is the stiffness measured experimentally from the slope of unloading curve. Based on known S and A values, E_r can be calculated. E_r can be used to find the true modulus of the material by the relation

$$\frac{1}{E_r} = \frac{\left(1-v^2\right)}{E} + \frac{\left(1-v_i^2\right)}{E_i}.$$ (3.4)

Here, E and v are Young's modulus and Poisson's ratio of the specimen under test. E_i and v_i are Young's modulus and Poisson's ratio of the indenter.

3.1.2   Experimental Setup

Experiments were performed at room temperature in a multimodule mechanical tester (NanoTest, Micro Materials Ltd., platform 2) [3, 4, 41], shown in Fig. 3.2. The instrument consists of a vertical pendulum pivoted on a frictionless spring. An indenter is attached to the pendulum that indents sample horizontally. The force on pendulum was applied through magnetic coils which allows for a very high sensitivity during experiments. The depth of indents during experiments was measured by capacitor plates located behind the indenter. Before experiments, calibrations were performed to get accurate measurements of load and depth. Depth calibration, load calibration, and frame compliance experiments were performed before conducting actual experiments. Tip radius of the Berkovich indenter was 20 nm.

During indents, samples were mounted on stage firmly to avoid any movement during experiments. The indentations were performed on different samples at the chosen surface as shown in Fig. 3.3. It was ensured that sample surface is free of defects before mounting samples. The indentation locations were chosen randomly and a series of indents at around ten points were performed. Experiments were performed on ten different samples to capture statistical property differences possibly

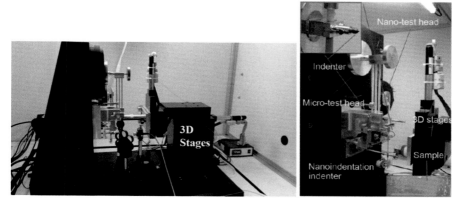

Fig. 3.2 Instrument setup, NanoTest, Micro Materials Ltd.

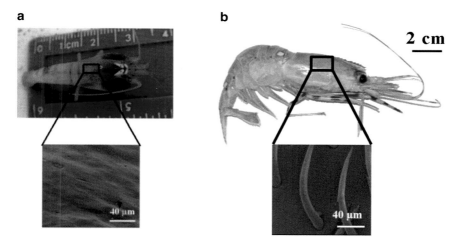

Fig. 3.3 Analyzed samples of (**a**) *Rimicaris exoculata* and (**b**) *Pandalus platyceros*

due to biological changes in samples. The experiments were performed at room temperatures of 25 °C. The nanoindentation used for the experiments incorporates a feedback mechanism to accurately stabilize temperature during experiments. It is always a challenge to simulate exactly the same experimental condition while performing experiments for long time intervals. In order to avoid discrepancies in collected data, the best approach used by most researchers is to do all experiments using identical experimental setup. In present experiments, all measurements were performed using same indenter, mounting technique, and samples at given temperatures. This gives us much more confidence in our data for comparison purpose. For each data point, around 50 indentations were performed.

3.1.3 Scanning Electron Microscopy (SEM) and Energy-Dispersive X-Ray (EDX)

SEM images were obtained by FEI Nova nanoSEM. All samples were coated with platinum layer of 10 nm. It is important to coat biosamples with a metallic layer to avoid charge accumulation on sample surface to discharge the accumulated charge from the surface of sample to have a clear image. The working distance for Nova SEM was 5 mm with 5.00 kV accelerating voltage in high-vacuum chamber. EDX analysis was performed using FEI Quanta 3D FEG dual beam SEM. It is hard to get enough counts per second for biological samples while performing elemental analysis at normal settings used for metals. Therefore, SEM aperture was increased to get higher counts per second for collection of elemental spectrum. The working distance for Quanta SEM was 10 mm with 20.00 kV accelerating voltage in high-vacuum chamber and cryogenic environment.

3.1.4 Sample Preparation

Shrimp samples of *Rimicaris exoculata* were provided by Dr. Juliette Ravaux from the mission Momardream in 2007 from depth of 2300 m. The shrimps were maintained for 10 h at 10 °C with 1 h heat shock at 30 °C in pressurized aquaria on board. These shrimp samples were further stored in liquid nitrogen in our lab. *Pandalus platyceros* shrimps were obtained in fresh unfrozen condition to prepare the samples. Shrimps were immediately stored in frozen condition after procurement. Samples for mechanical testing were prepared from these shrimps specimens. Figure 3.3 shows whole shrimps and prepared samples for *Rimicaris exoculata* and *Pandalus platyceros*. To prepare samples, shrimps were taken out and carapace was removed carefully without mechanically straining or producing any fracture in it. Samples were cut from carapace. Carapace is the exoskeleton of shrimp that protects cephalothoracic region of the shrimp. Cephalothoracic is the front part of shrimp that contains vital parts of shrimp including the mouth, eyes, antennas, etc. Carapace will be referred to as shrimp exoskeleton further in this paper. Shrimp exoskeletons were then dried in open for 3 days to prepare samples for indentation tests.

3.1.5 Substrate Effect

Substrate could play a major role in the measurement of reduced modulus values of thin samples. The general rule is that indentation depth should be less than 10 % of the total thickness of the sample [42–44]. The thickness of *Rimicaris exoculata* samples was 20 ± 1 μm and indentation depth was 300 nm, and the thickness of

Pandalus platyceros samples was $110 \pm 10\,\mu m$ and indentation depth was 1 μm which satisfies the 10 % rule. This verifies that substrate effect has been taken into account during indentations.

3.2   Results

Two different shrimp species *Rimicaris exoculata* and *Pandalus platyceros* were analyzed during the current study. Reduced modulus and hardness were measured by nanoindentation. There was a clear difference in the reduced modulus and hardness properties. This difference is because of the difference in the design of the shrimp exoskeleton. As shown in Figs. 3.4 and 3.5, there is a clear nonuniformity in stacking density and thickness of layers across cross section. It shows the inner side of exoskeleton at the bottom of the figure and outer side of exoskeleton at the top of the figure. Each visible layer is a layer of woven α-chitin–protein fiber network,

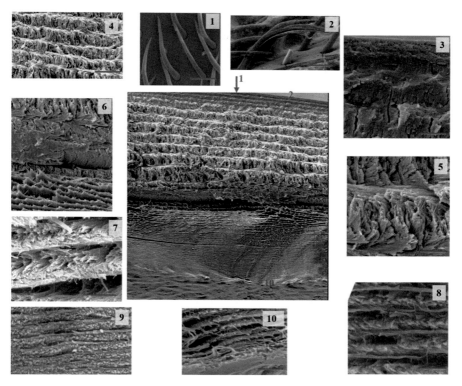

Fig. 3.4   Full cross section in the middle and zoomed sections of the cross section labeled from no. 1 to 10 around it for *Pandalus platyceros*. Scale bars, full cross section—20 μm, (*1*) 100 μm, (*2*) 50 μm, (*3*) 2 μm, (*4*) 10 μm, (*5*) 3 μm, (*6*) 5 μm, (*7*) 1 μm, (*8*) 500 nm, (*9*) 500 nm, (*10*) 500 nm

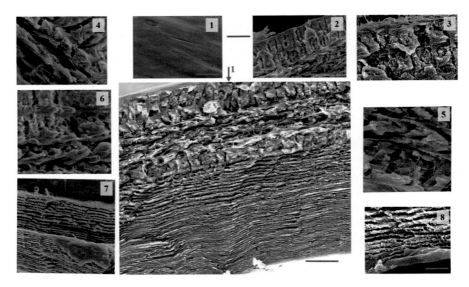

Fig. 3.5 Full cross section in the middle and zoomed sections of the cross section labeled from no. 1 to 8 around it for *Rimicaris exoculata*. Scale bars, full cross section—5 µm, (*1*) 20 µm, (*2*) 2 µm, (*3*) 1 µm, (*4*) 1 µm, (*5*) 1 µm, (*6*) 1 µm, (*7*) 2 µm, (*8*) 500 nm

the characteristic feature of exoskeleton of crustaceans [1, 14, 15]. It consists of different kinds of proteins and minerals. Both shrimp exoskeletons do have the similar features such as Bouligand structure and layered structure with thicker layers on the top and thinner layers at the bottom that can be visualized from Figs. 3.4 and 3.5. There is a difference in the length scale though. *Pandalus platyceros* in Fig. 3.4 shows thicker layers and overall higher thickness as compared to the *Rimicaris exoculata* shown in Fig. 3.5. Figure 3.4 also shows zoomed images from different parts of the cross section of *Pandalus platyceros* labeled from 1 to 10 on the full cross section. The surface of the exoskeleton shows hairy features (Fig. 3.4(1, 2)). The topmost layers are more closely knit together making them a fairly homogeneous structure (Fig. 3.4(3)). Different layers just below the top surface are visible (Fig. 3.4(4)). A close-up image on a layer shows the Bouligand pattern (Fig. 3.4(5)). The layers in this region of the shrimp are thicker as compared to the lower part. As we go down, we observe around the middle section that the thickness of layers changes to almost half of the original thickness (Fig. 3.4(6)). After this point, the layer thickness keeps on decreasing (Fig. 3.4(7–9)) with the minimum thickness layers at the bottom (Fig. 3.4(10)) of the cross section of exoskeleton.

Similarly, Fig. 3.5 shows zoomed images from different parts of the cross section of *Rimicaris exoculata* labeled from 1 to 8 on the full cross section. The surface of the exoskeleton shows no hairs (Fig. 3.5(1)). The topmost layers (Fig. 3.5(2)) are not as homogeneous as we observed in Fig. 3.4. Different layers just below the top surface are visible in Fig. 3.5(3, 4). A close-up image on a layer shows the Bouligand pattern (Fig. 3.5(3)). The layers in this region of the shrimp are thicker as compared to the

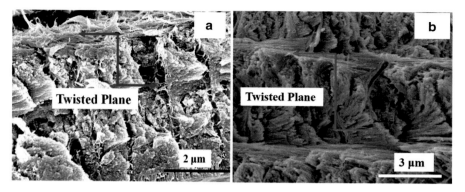

Fig. 3.6 Bouligand structure of shrimp exoskeleton. (**a**) *Rimicaris exoculata.* (**b**) *Pandalus platyceros*

lower part. The thickness of the layers is not changing in a uniform fashion though. The topmost layers are thicker (Fig. 3.5(2, 3)) following the layers of lower thickness (Fig. 3.5(4)) that further increase in thickness (Fig. 3.5(5)) till the half of the total thickness of cross section. As we go down, we observe around the middle section that the thickness of layers changes again to much thinner layers (Fig. 3.5(6)). After this point, the layer thickness keeps on decreasing (Fig. 3.5(7)) with the minimum thickness layers at the bottom (Fig. 3.5(8)) of the cross section of exoskeleton.

The maximum thickness of a layer in the *Pandalus platyceros* is 2 μm as opposed to 1 μm in *Rimicaris exoculata* exoskeleton. *Rimicaris exoculata* shows a nonuniform change on the thickness of the layers as compared to *Pandalus platyceros*. The minimum thickness at the bottom most regions is in the range of 20–30 nm for *Rimicaris exoculata*, while in *Pandalus platyceros* it is in the range of 50–100 nm. The similarity in the individual layers of both shrimps can be exhibited from the Bouligand pattern as shown in Fig. 3.6.

Further analysis of the SEM images shows the failure patterns in both species exoskeletons. Figure 3.7(1) shows the layers of cross section in *Pandalus platyceros* upper section with bend layers in Fig. 3.7(3). A closer picture showing the delamination of layers is shown in Fig. 3.7(2). It shows the cross-linked pattern between two layers at the edges of two layers. These interconnected layers enhance the flexibility and robustness of these exoskeletons. Similar delamination in the *Rimicaris exoculata* exoskeleton can be seen from Fig. 3.8(1–2). In this exoskeleton, the individual layers are more closely packed. The layers are much thinner with fracture of layers accompanied by fiber out (Fig. 3.4(3–4)). These layers show bending at even smaller length scales (Fig. 3.8(2)).

In addition to the difference in the structure, these species also show difference in the chemical composition of the exoskeleton. Chemical composition at the top layers at the sites of indention was compared by the taking EDX spectra. Both spectra were collected using same Quanta 3D SEM at Purdue University. The elemental composition calculated from the spectra is given in Table 3.1.

Fig. 3.7 Images showing failures and fractures in the cross section of *Pandalus platyceros* (*1*) in top layer, (*2*) delamination, and (*3*) bending of layers around hairs. Scale bars, (*1*) 15 µm, (*2*) 6 µm, (*3*) 10 µm

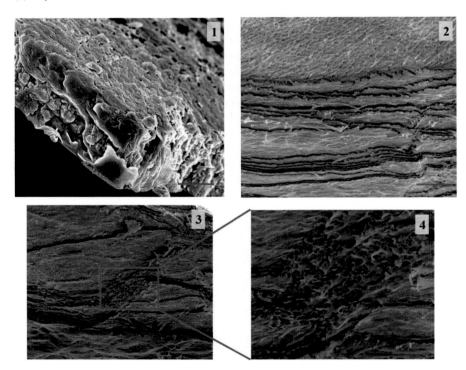

Fig. 3.8 Images showing failures and fractures in the cross section of *Pandalus platyceros* (*1*) delaminated layer, (*2*) interlayer delamination, (*3*) fracture in the layers, and (*4*) zoomed image of fiber pullout from fracture in layers. Scale bars, (*1*) 3 µm, (*2*) 5 µm, (*3*) 10 µm, (*4*) 3 µm

Table 3.1 Comparison of quantitative data by EDX analysis on cross section of shrimp exoskeleton for *Rimicaris exoculata* and *Pandalus platyceros*

Element	*Rimicaris exoculata* (at.%)	*Pandalus platyceros* (at.%)
C	42.89±0.97	46.20±1.19
O	27.76±1.89	42.41±1.79
P	3.14±1.12	–
Ca	26.2±1.73	9.86±2.20

Table 3.2 Reduced modulus and hardness of shrimp exoskeleton for *Rimicaris exoculata* and *Pandalus platyceros*

Property	*Rimicaris exoculata*	*Pandalus platyceros*
Reduced modulus	8.27±0.89	27.38±2.3
Hardness	0.31±0.07	1.52±0.16

This is comparable to the chemical composition obtained by several researchers on coastal shrimps [45–51]. *Rimicaris exoculata* shows some variations in composition with noticeable phosphorus. Table 3.1 gives relative quantitative analysis of individual elements present in the exoskeleton of both shrimps. The presence of Ca supports the fact that *Rimicaris exoculata* survives at very high pressures and needs to have more structural strength to survive. Phosphorus is one of the materials from volcanic vents in nearby habitat of *Rimicaris exoculata*. It also became part of shrimp exoskeleton with evolution of shrimps near deep-sea vents possibly via food sources. A combination of observations from Figs. 3.4 to 3.8 and Table 3.1 reveals that *Rimicaris* exoskeleton has higher content of mineral and reduced protein layer thicknesses in comparison to *Pandalus platyceros*. The reduced modulus and hardness of both species are given in Table 3.2. The difference in these properties can be attributed to the changes at the structural and chemical composition as mentioned in earlier discussion. Figure 3.4(3) shows closely packed layers in *Pandalus platyceros* at the top, while Fig. 3.5(2) indicates that layers on the top of *Rimicaris exoculata* show some porosity. This is also one reason for the reduced mechanical properties.

Further analysis was performed on the creep data obtained from the nanoindentation experiments. A hold period was applied at the maximum load to collect dwell data. Hardness of a material is its ability to resist deformation under applied stress. In the present material, creep compliance follows the trend of hardness. There are several formulations to fit the nanoindentation creep given by Oyen et al. [52–54] and Lu et al. [55]. The model chosen in the current paper is the most fit for the nanoindentation data of the creep. It uses the dwell data over time for the holding period to calculate creep compliance parameters. In order to model the load-deformation behavior exhibited by examined shrimp exoskeletons, the creep data is fitted with the creep function given for viscoelastic materials by Oyen [52] as

$$h^2 = \frac{\gamma^2}{\pi \tan \psi} \left\{ C_o k t_R - \sum C_i k \tau_i \exp\left(\frac{-t}{\tau_i}\right) \left[\exp\left(\frac{t_R}{\tau_i}\right) - 1 \right] \right\}. \qquad (3.5)$$

Table 3.3 Creep compliance fitting parameters from Eq. 3.5 on experimental dwell data

Species	C_o (Pa^{-1})	C_1 (Pa^{-1})	τ_1 (s)
Rimicaris exoculata	1.45e−6±8.21e−8	4.60e−7±4.97e−9	86.33±1.24
Pandalus platyceros	1.08e−6±8.16e−8	1.38e−7±4.80e−9	68.00±1.63

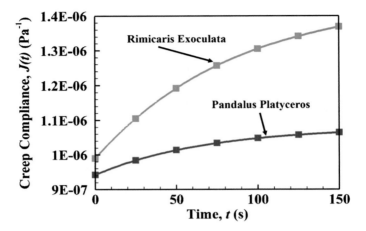

Fig. 3.9 Creep compliance functions $J(t)$ for *Rimicaris exoculata* and *Pandalus platyceros* from Eq. 3.6

Here, h is the indentation depth during dwell period, ψ is the indenter included angle, γ is a constant relating contact displacement to total displacement, k is the loading rate, t_R is the start time of dwell period, C_i and τ_i are the creep compliance coefficients, and t is the time period of loading. One term approximation of Eq. (3.5) is fitted to the experimental creep data to find coefficients C_i and τ_i. Respective coefficients for both *Rimicaris exoculata* and *Pandalus platyceros* are tabulated in Table 3.3.

The creep compliance parameters obtained from Eq. 3.5 were then used to plot the creep compliance given in Fig. 3.9 from Eq. 3.6:

$$J(t) = C_0 - \sum_{i=1}^{j} C_i \exp\left(\frac{-t}{\tau_i}\right). \tag{3.6}$$

Here, C_i and τ_i are the creep compliance coefficients, and t is the time period of loading. As shown in Fig. 3.9, the creep compliance for *Rimicaris exoculata* is higher as compared to *Pandalus platyceros* because of the lower hardness of *Rimicaris exoculata* exoskeleton. Basic building blocks of the *Rimicaris exoculata* and the *Pandalus platyceros* exoskeleton are polymer chains made up of chitin proteins and calcium-based biominerals. Matrix material is made up of chitin–protein fibers woven together in a planar structure. Spacing between woven fibers is filled with biominerals, mostly of $CaCO_3$. These layers are then arranged to form twisted plywood structure known as Bouligand pattern [1, 14]. Such hierarchy in this structure makes it similar to polymer composites. The creep compliance functions fit nicely to the creep data of the shrimp exoskeletons.

The other reasons for the changes in structure that lead to changes in the mechanical properties can be found in the habitat of both species. *Rimicaris exoculata* lives at high temperature and pressure at depths of 2300 m and near high-temperature deep-sea vents [56–60] (~500 °C). *Pandalus platyceros* lives at sea level and does not need to sustain high pressures and temperatures. *Rimicaris exoculata* has higher mineral content for toughness but maintains porosity in the structural design to keep its structure flexible. At sea level, *Pandalus platyceros* needs to encounter several animals that are above it in the food chain, so it needs to maintain higher structural strength and thus have higher modulus and hardness. These two species provide a good example that exhibits changes in its microstructure to develop its exoskeleton according to the requirements of their habitats.

3.3   Conclusion

A series of experiments were performed at room temperature on the exoskeletons of *Rimicaris exoculata* and *Pandalus platyceros* in order to understand their structural dependent mechanical strength. The structures were also studied extensively using the SEM (scanning electron microscopy) and EDX (energy-dispersive X-ray) analyses. A comparison is drawn between the properties of exoskeletons of *Rimicaris exoculata* and *Pandalus platyceros* based on their microstructure and chemical composition. The difference in the reduced modulus and hardness for both shrimp species is attributed to their habitat and subsequent changes in the microstructural design such as change in the individual layer thickness, pattern of layer thickness, and chemical composition. The results are also fitted with viscoelastic creep compliance functions which further show *Rimicaris* to have higher creep as compared to *Pandalus platyceros* due to its lower hardness.

Acknowledgments The authors express their sincere thanks to Dr. Juliette Ravaux, Université Pierre et Marie Curie, for providing samples of *Rimicaris exoculata*. Also, they would like to acknowledge the excellent technical assistance of Dr. Christopher J. Gilpin, Chia-Ping Huang, and Laurie Mueller with scanning electron microscopy and energy-dispersive X-ray at Purdue University. Lastly I would like to thank my colleagues Dr. Devendra Dubey, Dr. Ming Gan, Dr. You Sung Han, and Dr. Hongsuk Lee for helpful discussions.

References

1. D. Raabe, C. Sachs, P. Romano, The crustacean exoskeleton as an example of a structurally and mechanically graded biological nanocomposite material. Acta Mater. **53**(15), 4281 (2005)
2. S. Nikolov, M. Petrov, L. Lymperakis, M. Friák, C. Sachs, H.-O. Fabritius, D. Raabe, J. Neugebauer, Revealing the design principles of high-performance biological composites

using ab initio and multiscale simulations: the example of lobster cuticle. Adv. Mater. **22**(4), 519 (2010)

3. D. Verma, V. Tomar, Structural-nanomechanical property correlation of shallow water shrimp (*Pandalus platyceros*) exoskeleton at elevated temperature. J. Bionic Eng. **11**(3), 360 (2014)

4. D. Verma, V. Tomar, An investigation into environment dependent nanomechanical properties of shallow water shrimp (Pandalus platyceros) exoskeleton. Mater. Sci. Eng. C **44**, 371 (2014)

5. F. Bouville, E. Maire, S. Meille, B. Van de Moortèle, A.J. Stevenson, S. Deville, Strong, tough and stiff bioinspired ceramics from brittle constituents. Nat. Mater. **13**(5), 508 (2014)

6. J.-y. Sun, J. Tong, Fracture toughness properties of three different biomaterials measured by nanoindentation. J. Bionic Eng. **4**(1), 11 (2007)

7. S. Bechtle, S.F. Ang, G.A. Schneider, On the mechanical properties of hierarchically structured biological materials. Biomaterials **31**(25), 6378 (2010)

8. G. Mayer, New toughening concepts for ceramic composites from rigid natural materials. J. Mech. Behav. Biomed. Mater. **4**(5), 670 (2011)

9. M.S. Wu, Strategies and challenges for the mechanical modeling of biological and bio-inspired materials. Mater. Sci. Eng. C **31**(6), 1209 (2011)

10. P.-Y. Chen, A.Y.-M. Lin, J. McKittrick, M.A. Meyers, Structure and mechanical properties of crab exoskeletons. Acta Biomater. **4**(3), 587 (2008)

11. J. Lian, J. Wang, Microstructure and Mechanical Properties of Dungeness Crab Exoskeletons, in *Mechanics of Biological Systems and Materials*, ed. by T. Proulx, vol. 2 (Springer, New York, 2011), p. 93

12. H.R. Hepburn, I. Joffe, N. Green, K.J. Nelson, Mechanical properties of a crab shell. Comp. Biochem. Physiol. A Physiol. **50**(3), 551 (1975)

13. M.M. Giraud-Guille, Fine structure of the chitin-protein system in the crab cuticle. Tissue Cell **16**(1), 75 (1984)

14. D. Raabe, P. Romano, C. Sachs, H. Fabritius, A. Al-Sawalmih, S.-B. Yi, G. Servos, H. Hartwig, Microstructure and crystallographic texture of the chitin–protein network in the biological composite material of the exoskeleton of the lobster *Homarus americanus*. Mater. Sci. Eng. A **421**(1), 143 (2006)

15. F. Boßelmann, P. Romano, H. Fabritius, D. Raabe, M. Epple, The composition of the exoskeleton of two crustacea: the American lobster Homarus americanus and the edible crab Cancer pagurus. Thermochim. Acta **463**(1–2), 65 (2007)

16. L. Wang, J. Song, C. Ortiz, M.C. Boyce, Anisotropic design of a multilayered biological exoskeleton. J. Mater. Res. **24**(12), 3477 (2009)

17. Y. Seki, M.S. Schneider, M.A. Meyers, Structure and mechanical behavior of a toucan beak. Acta Mater. **53**(20), 5281 (2005)

18. Y. Seki, B. Kad, D. Benson, M.A. Meyers, The toucan beak: structure and mechanical response. Mater. Sci. Eng. C **26**(8), 1412 (2006)

19. G. Mayer, New classes of tough composite materials—lessons from natural rigid biological systems. Mater. Sci. Eng. C **26**(8), 1261 (2006)

20. X. Li, P. Nardi, Micro/nanomechanical characterization of a natural nanocomposite material—the shell of Pectinidae. Nanotechnology **15**(1), 211 (2004)

21. B. Ji, H. Gao, Mechanical properties of nanostructure of biological materials. J. Mech. Phys. Solid **52**(9), 1963 (2004)

22. B. Chen, X. Peng, J.G. Wang, X. Wu, Laminated microstructure of Bivalva shell and research of biomimetic ceramic/polymer composite. Ceram. Int. **30**(7), 2011 (2004)

23. H. Gao, B. Ji, I.L. Jäger, E. Arzt, P. Fratzl, Materials become insensitive to flaws at nanoscale: lessons from nature. Proc. Natl. Acad. Sci. **100**(10), 5597 (2003)

24. C.-a. Wang, Y. Huang, Q. Zan, H. Guo, S. Cai, Biomimetic structure design—a possible approach to change the brittleness of ceramics in nature. Mater. Sci. Eng. C **11**(1), 9 (2000)

25. Q.L. Feng, F.Z. Cui, G. Pu, R.Z. Wang, H.D. Li, Crystal orientation, toughening mechanisms and a mimic of nacre. Mater. Sci. Eng. C **11**(1), 19 (2000)

26. F. Barthelat, J.E. Rim, H.D. Espinosa, in *Applied Scanning Probe Methods XIII*. A Review on the Structure and Mechanical Properties of Mollusk Shells–Perspectives on Synthetic Biomimetic Materials. (Springer, New York, 2009), p. 17

27. G.M. Luz, J.F. Mano, Biomimetic design of materials and biomaterials inspired by the structure of nacre. Philos. Trans. R. Soc. A Math. Phys. Eng. Sci. **367**(1893), 1587 (2009)
28. S. Raman, R. Kumar, Construction and nanomechanical properties of the exoskeleton of the barnacle, *Amphibalanus reticulatus*. J. Struct. Biol. **176**(3), 360 (2011)
29. Y. Bouligand, Twisted fibrous arrangements in biological materials and cholesteric mesophases. Tissue Cell **4**(2), 189 (1972)
30. C.A. Melnick, Z. Chen, J.J. Mecholsky, Hardness and toughness of exoskeleton material in the stone crab, Menippe mercenaria. J. Mater. Res. **11**(11), 2903 (1996)
31. B.-S. Kim, D.J. Mooney, Development of biocompatible synthetic extracellular matrices for tissue engineering. Trends Biotechnol. **16**(5), 224 (1998)
32. H. Shin, S. Jo, A.G. Mikos, Biomimetic materials for tissue engineering. Biomaterials **24**(24), 4353 (2003)
33. J. Chen, Q.-Q. Ni, Y. Xu, M. Iwamoto, Lightweight composite structures in the forewings of beetles. Compos. Struct. **79**(3), 331 (2007)
34. J. Chen, C. Gu, S. Guo, C. Wan, X. Wang, J. Xie, X. Hu, Integrated honeycomb technology motivated by the structure of beetle forewings. Mater. Sci. Eng. C **32**(7), 1813 (2012)
35. J. Chen, J. Xie, H. Zhu, S. Guan, G. Wu, M.N. Noori, S. Guo, Integrated honeycomb structure of a beetle forewing and its imitation. Mater. Sci. Eng. C **32**(3), 613 (2012)
36. J. Chen, G. Wu, Beetle forewings: epitome of the optimal design for lightweight composite materials. Carbohydr. Polym. **91**(2), 659 (2013)
37. S. Stankovich, D.A. Dikin, G.H.B. Dommett, K.M. Kohlhaas, E.J. Zimney, E.A. Stach, R.D. Piner, S.T. Nguyen, R.S. Ruoff, Graphene-based composite materials. Nature **442**(7100), 282 (2006)
38. T. Ramanathan, A.A. Abdala, S. Stankovich, D.A. Dikin, M. Herrera Alonso, R.D. Piner, D.H. Adamson, H.C. Schniepp, X. Chen, R.S. Ruoff, S.T. Nguyen, I.A. Prud'Homme, R.K. Aksay, L.C. Brinson, Functionalized graphene sheets for polymer nanocomposites. Nat. Nanotechnol. **3**(6), 327 (2008)
39. W.C. Oliver, G.M. Pharr, Improved technique for determining hardness and elastic modulus using load and displacement sensing indentation experiments. J. Mater. Res. **7**(6), 1564 (1992)
40. G. Pharr, Measurement of mechanical properties by ultra-low load indentation. Mater. Sci. Eng. A **253**(1), 151 (1998)
41. M. Gan, V. Tomar, Role of length scale and temperature in indentation induced creep behavior of polymer derived Si–C–O ceramics. Mater. Sci. Eng. A **527**(29–30), 7615 (2010)
42. R. Saha, W.D. Nix, Effects of the substrate on the determination of thin film mechanical properties by nanoindentation. Acta Mater. **50**(1), 23 (2002)
43. C. Gamonpilas, E.P. Busso, On the effect of substrate properties on the indentation behaviour of coated systems. Mater. Sci. Eng. A **380**(1–2), 52 (2004)
44. D. Kramer, A. Volinsky, N. Moody, W. Gerberich, Substrate effects on indentation plastic zone development in thin soft films. J. Mater. Res. **16**(11), 3150 (2001)
45. S. Ravichandran, G. Rameshkumar, A.R. Prince, Biochemical composition of shell and flesh of the Indian white shrimp Penaeus indicus (H. milne Edwards 1837). Am.-Euras. J. Sci. Res. **4**(3), 191 (2009)
46. F. Ehigiator, E. Oterai, Chemical composition and amino acid profile of a Caridean prawn (Macrobrachium vollenhovenii) from Ovia river and tropical periwinkle (Tympanotonus fuscatus) from Benin River, Edo State, Nigeria. Int. J. Res. Rev. Appl. Sci. **11**(1), 162–167 (2012)
47. I.A. Emmanuel, H.O. Adubiaro, O.J. Awodola, Comparability of chemical composition and functional properties of shell and flesh of Penaeus notabilis. Pak. J. Nutr. **7**(6), 741 (2008)
48. R.H. Rødde, A. Einbu, K.M. Vårum, A seasonal study of the chemical composition and chitin quality of shrimp shells obtained from northern shrimp (Pandalus borealis). Carbohydr. Polym. **71**(3), 388 (2008)
49. H.M. Ibrahim, M.F. Salama, H.A. El-Banna, Shrimp's waste: chemical composition, nutritional value and utilization. Food/Nahrung **43**(6), 418 (1999)
50. F. Shahidi, J. Synowiecki, Isolation and characterization of nutrients and value-added products from snow crab (Chionoecetes opilio) and shrimp (Pandalus borealis) processing discards. J. Agric. Food Chem. **39**(8), 1527 (1991)

51. M. Islam, S. Masum, M. Rahman, M. Moll, A. Shaikh, S. Roy, Preparation of chitosan from shrimp shell and investigation of its properties. Int. J. Basic Appl. Sci. **11**(1), 116 (2011)
52. M. Oyen, Analytical techniques for indentation of viscoelastic materials. Philos. Mag. **86**(33–35), 5625 (2006)
53. M.L. Oyen, R.F. Cook, Load–displacement behavior during sharp indentation of viscous–elastic–plastic materials. J. Mater. Res. **18**(01), 139 (2003)
54. R.F. Cook, M.L. Oyen, Nanoindentation behavior and mechanical properties measurement of polymeric materials. Int. J. Mater. Res. **98**(5), 370 (2007)
55. H. Lu, B. Wang, J. Ma, G. Huang, H. Viswanathan, Measurement of creep compliance of solid polymers by nanoindentation. Mech. Time-Depend. Mater. **7**(3–4), 189 (2003)
56. H. Fricke, O. Giere, K. Stetter, G.A. Alfredsson, J.K. Kristjansson, P. Stoffers, J. Svavarsson, Hydrothermal vent communities at the shallow subpolar Mid-Atlantic ridge. Mar. Biol. **102**(3), 425 (1989)
57. D. Desbruyères, M. Biscoito, J.C. Caprais, A. Colaço, T. Comtet, P. Crassous, Y. Fouquet, A. Khripounoff, N. Le Bris, K. Olu, R. Riso, P.M. Sarradin, M. Segonzac, A. Vangriesheim, Variations in deep-sea hydrothermal vent communities on the Mid-Atlantic Ridge near the Azores plateau. Deep-Sea Res. I Oceanogr. Res. Pap. **48**(5), 1325 (2001)
58. S.T. Ahyong, New species and new records of hydrothermal vent shrimps from New Zealand (Caridea: Alvinocarididae, Hippolytidae). Crustaceana **82**(7), 775 (2009)
59. R.C. Vrijenhoek, Genetic diversity and connectivity of deep-sea hydrothermal vent metapopulations. Mol. Ecol. **19**(20), 4391 (2010)
60. D.P. Connelly, J.T. Copley, B.J. Murton, K. Stansfield, P.A. Tyler, C.R. German, C.L. Van Dover, D. Amon, M. Furlong, N. Grindlay, N. Hayman, V. Huhnerbach, M. Judge, T. Le Bas, S. McPhail, A. Meier, K.-i. Nakamura, V. Nye, M. Pebody, R.B. Pedersen, S. Plouviez, C. Sands, R.C. Searle, P. Stevenson, S. Taws, S. Wilcox, Hydrothermal vent fields and chemosynthetic biota on the world's deepest seafloor spreading centre. Nat. Commun. **3**, 620 (2012)

Chapter 4
Molecular Modeling: A Review of Nanomechanics Based on Molecular Modeling

Abstract Nature's design and engineering of biological material systems have always intrigued researchers for their extraordinary properties and structure–property–function relationships. One aspect of biomaterials science and engineering is to understand the underlying mechanisms, design, and fabrication pathways of such biological materials, which will have benefit in multiple disciplines such as prosthetic implants, regenerative medicine, self-healing materials, novel high-strength biomimetic materials, and bioenergy applications. The focus of this review is on the chemo-mechanics of the organic–inorganic interfaces and its correlation with overall mechanical behavior. This understanding is vital for selecting appropriate constituents, their size scales and their relative arrangements, which in turn is governed by the functional requirements of the composite materials.

Keywords Atomic modeling • Nanoscale modeling • Effect of interfaces • Interface chemistry

4.1 Introduction

Nature's design and engineering of biological material systems have always intrigued researchers for their extraordinary properties and structure–property–function relationships. The toughness of spider silk, the strength and lightweight of bamboos, self-healing of bone, high toughness of nacre, and the adhesion abilities of the gecko's feet are a few of the many examples of high-performance natural materials. Biological materials usually have excellent multifunctional properties. For example, bone provides structural support as well as serves as home for blood cell production. The extraordinary adhesive capabilities of gravity defying gecko's feet have left researchers wondering for centuries [1, 2]. Biological materials science and engineering has emerged as a young and vibrant discipline since its inception in the 1990s. One aspect of biomaterials science and engineering is to understand the underlying mechanisms, design, and fabrication pathways of such biological materials, which will have benefit in multiple disciplines such as prosthetic implants, regenerative medicine, self-healing materials, novel high-strength biomimetic materials, and bioenergy applications. The focus of this review is on hard biomaterials. In most fail safe designs, engineering materials are limited by

© Springer Science+Business Media New York 2015
V. Tomar et al., *Multiscale Characterization of Biological Systems*,
DOI 10.1007/978-1-4939-3453-9_4

their low resistance to fracture or low fracture toughness. From the fracture mechanics perspective, hard biological materials easily outperform engineering materials in developing fracture resistance during crack growth. This is attributed to the ability of their microstructure to develop toughening mechanisms acting either ahead or behind the crack tip.

Generally, hard biomaterials have multilevel hierarchical structure with an organic phase and an inorganic phase blended together at the fundamental length scales. Hard biological materials such as bone, nacre, dentin, and wood, etc., are not only lightweight but also possess high toughness and strength. Nacre is nearly 3000 times more fracture resistant than its constituent mineral [3]. The concept that in order to explain the material properties of compact cortical bone, it was necessary to treat it as a complex structural hierarchical composite was introduced by [4]. The necessity of looking at bone as a hierarchical structural composite was originally introduced by [5]. Materials such as bone and nacre have such multilevel hierarchical structural design that concept of stress concentration at flaws remains invalid, leading to flaw tolerant structure [6]. Such biological materials have been reviewed in appreciable detail, in the context of their hierarchical structure, material properties, and failure mechanisms [7–10]. Attractive mechanical properties of hard biological materials have given birth to biomimetism and new biomaterials which have also been reviewed by many [10–19].

An important aspect to focus in biomaterials engineering of hard biomaterials is the chemo-mechanics of the organic–inorganic interfaces and its correlation with overall mechanical behavior. This understanding is vital for selecting appropriate constituents, their size scales and their relative arrangements, which in turn is governed by the functional requirements of the composite materials. For example, three-dimensional explicit atomistic failure analyses of model tropocollagen (TC)–hydroxyapatite (HAP) interfacial biomaterial (similar to material found in bone tissues) performed at the nanoscopic length scale [20–22] have pointed out that maximizing the contact area between the TC and HAP phases results in higher interfacial strength as well as higher fracture strength. Analyses have also shown that high toughness and strain-hardening behavior of such biomaterial is due to reconstitution of columbic interactions between TC and HAP surfaces during interfacial sliding due to mechanical deformation. In this focused review, an analysis of work performed in understanding interfacial nanomechanics of hard biomaterials is presented. The focus is on presenting an understanding that could be directed for use in biomedical engineering and biomaterial development.

4.2 Bioengineering and Biomimetics

It can be argued that all materials are hierarchically structured as the change in length scale affects the mechanisms of deformation and damage. However, in biological materials, this hierarchical organization is inherent to the design. Unlike engineered synthetic materials, the structural design and material property in biological materials are intricately coupled with each other for carrying out required functions optimally. The most prominent feature of such hierarchical design is the organic–inorganic interfacial interaction and the resulting biomaterial mechanics. To understand the importance of such interfacial interactions, the next section describes the work done in two most widely studied biological materials, bone and nacre.

4.2.1   Bone

Bone is an excellent example of stiff biological nanocomposite materials. From the mechanical strength viewpoint, its main constituents are TC molecules (collagen) and biological HAP mineral. Biological HAP is a poorly crystalline impure form of the compound HAP (chemical formula: $Ca_5(PO_4)_3OH$), containing constituents such as carbonate, citrate, magnesium, fluoride, and strontium [23, 24]. Despite the fact that TC is a soft phase and HAP is brittle, together they form a material of high mechanical strength and fracture toughness [25]. The structure of bone is organized over several length scales, with six to seven levels of hierarchy [9, 26]. Figure 4.1 shows the several hierarchical features from macroscopic to atomic scale. TC molecules assemble into collagen fibrils in a hydrated environment, which mineralize by formation of HAP crystals in the gap regions that exist due to the staggered arrangement. These mineralized collagen fibrils (MFCs) assemble together with extrafibrillar matrix to form the next hierarchical layer of bone. While the structures at scales larger than MCFs vary for different bone types, MFCs are highly conserved, nanostructural primary building blocks of bone that are found universally [26, 29–31]. As shown in Fig. 4.1, at the fundamental scale, collagen fibrils are formed by staggered self-assembly of 300 nm long TC molecules, and mineral HAP occupies the gap regions along with water. Collagen fibril consists of some other constituents as well (such as non-collagenous proteins, glycans, mineral impurities, etc.). However, interfacial interactions between the TC and the HAP phases along with their hierarchical arrangement are thought to be the most important factors imparting high mechanical strength [32]. As shown using dotted and solid ellipses in Fig. 4.1, a staggered arrangement is responsible for possible tension–shear type of load transfer between TC molecules and HAP crystals in bone and in similar other biomaterials [33]. There have been growth and mineralization-based explanations behind the existence of such an arrangement [34–36]. However, it may also be possible that the existence of such structural arrangement is driven by its important role in imparting high mechanical strength.

The mechanical behavior of biological materials with a view to understand the role of TC molecules and HAP mineral has been earlier analyzed using experiments, modeling, and simulations. Experimental approaches have focused on analyzing tensile failure of single collagen fibers and fibrils [32, 37–40] and on analyzing tissue structural features at the nanoscale and its relation with the bone tissue failure [28, 41, 42]. Modeling using the continuum approaches has focused on understanding the role played by the shear strength of TC molecules and the tensile strength of

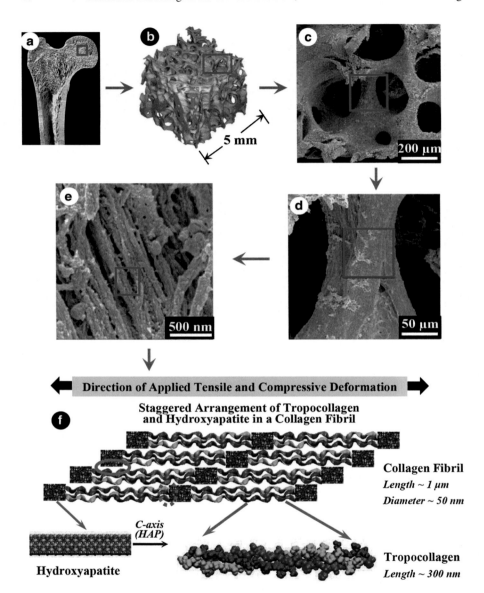

Fig. 4.1 A representation of the hierarchical structure in bone: (**a**) longitudinal cross section of end of the human femur showing the trabecular bone inside, (**b**) microstructure of trabecular bone resembling the honeycomb structure, (**c**) high-resolution view of the strut-like structures in microporous structure, (**d**) showing one of such struts called trabeculae with aligned mineralized collagen fibers, (**e**) fracture surface of human bone showing mineralized collagen fibrils, and (**f**) a schematic of staggered and layered assembly of tropocollagen (TC) molecules and hydroxyapatite (HAP) blocks to form a mineralized collagen fibril. Three different colors in TC molecule depict three polypeptide chains forming a triple helix. HAP crystal *c*-axis is along the loading direction. *Solid* and *dotted ellipses* in *pink* show two possible types of interactions between TC and HAP surface, one with TC perpendicular to HAP surface and other with TC parallel to HAP surface. Images in (**c**) and (**d**) are borrowed with permission from [27] and image (**e**) from [28]

HAP mineral in flaw tolerant hierarchical structural design of biomaterials [25, 29, 33, 43]. Explicit simulations using molecular dynamics (MD) schemes have focused on understanding mechanical behavior and properties of TC molecules in different structural configurations [44], on understanding hierarchical organization of TC molecules into collagen fibrils and its effect on mechanical properties [45], on understanding properties of hydrated TC molecules [46, 47], and on understanding TC molecule stability with respect to changes in residue sequences [48].

4.2.1.1 Nanoscopic Analytical Modeling of Bone and Bone-Like Materials

Several studies have been performed to understand mechanical behavior of bone and bone-like nanocomposite materials. Deformation, damage, and failure mechanisms at nanoscale seem to be as important as the mechanisms at macroscopic (beyond fibril size) length scale. Present work focuses on the insights gained at the fundamental nanoscopic length scales. One of the interesting studies performed by Gao and Ji proposes an analytical model called tension–shear chain model (TSC), Fig. 4.2, explaining the load transfer mechanism in generic hierarchical nanocomposite material systems such as bone and bone-like materials [6, 25, 33, 43]. Under uniaxial tension, the path of load transfer in the staggered nanostructure follows a tension–shear chain with mineral platelets under tension and the soft matrix under shear (Fig. 4.2b). The work has addressed the prevalence and importance of nanoscale features in biological composites and also developed methodology for estimating interfacial shear strength, fracture localization width, and characteristic length of mineral crystal (Fig. 4.3). A fractal-based analysis of TSC model is also performed to gain understanding for the need of different hierarchical structural levels (Fig. 4.3). The analysis shows that nanometer size of brittle mineral crystals is essential to make them insensitive to crack like flaws [6].

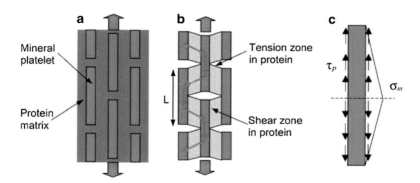

Fig. 4.2 Models of biocomposites. (**a**) Perfectly staggered mineral inclusions embedded in protein matrix. (**b**) A tension–shear chain model of biocomposites in which the tensile regions of protein are eliminated to emphasize the load transfer within the composite structure. (**c**) The free body diagram of a mineral crystal. Image is borrowed with permission from [25]

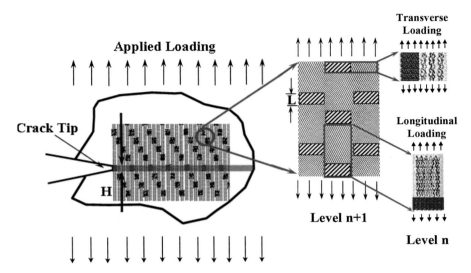

Fig. 4.3 Schematic representation of a crack, the fracture localization zone, and the hierarchical levels in the tension–shear chain model arrangement of hydroxyapatite mineral platelets and tropocollagen molecules in a tropocollagen–hydroxyapatite biocomposite

Among collagenous tissues such as the bone and tendon, the common structural feature is the mineralized collagen fibril acting as a building block. Accurate modeling of such mineralized collagen fibril is important and will require the right measurements of the dimensions, mechanical properties, and relative arrangement of the nano-length-scale constituents. Collagen molecules form ~100–200 nm diameter fibrils, with mineral particles inside and outside the surface [49]. Most studies describe the mineral particles to be plate shaped with a wider range of geometrical dimensions. The thickness of the platelets can range from 2 to 7 nm, the length from 15 to 200 nm, and the width from 10 to 80 nm [30, 31]. TC molecule has a length ~300 nm. Both HAP mineral and TC molecules are arranged in a staggered fashion (as shown in Fig. 4.1f) with a gap of ~67 nm between two successive TC molecules creating space for mineral nucleation and growth [26]. With the current state-of-the-art methods, complete measurements for anisotropic stress–strain–strength properties for biological HAP mineralite and TC molecule are not available. Without these measurements, current models generally utilize elastic moduli measured for HAP polycrystallites. This should not lead to an accurate modeling of mineralized fibril because the HAP mineral in calcified tissue is usually in the form of nanocrystals.

4.2.1.2   Atomistic Modeling TC–HAP Biomimetic Materials: Role of Interfaces and Structural Arrangement

In the structural studies of hard biological materials, it is observed that the mineral crystals tend to have a preferential alignment with respect to the loading direction and longitudinal axis of the polypeptide molecules [34, 35, 50, 51]. For example, in bone, both HAP crystal largest dimension (c-axis) and TC longitudinal axis are aligned along the maximum load-bearing direction [52], with a specific staggered structural arrangement (Figs. 4.1 and 4.4), permitting a maximum contact area. Resulting nanoscale interfacial interactions between TC phase and HAP phase are a strong determinant of the strength of such materials. Recent works by [20–22, 53, 54] present a mechanistic understanding of such interfacial interactions by examining idealized TC and HAP interfacial biomaterials. The HAP mineral modeled in this study is limited to the pure HAP as described in Sect. 4.2.1. Other impurities such as carbonate, citrate, magnesium, fluoride, etc., which are present in the biological HAP mineral can be incorporated into the HAP crystal lattice or absorbed

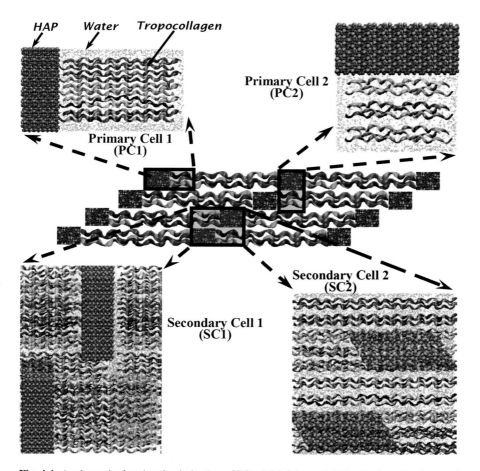

Fig. 4.4 A schematic showing the derivation of PC1, PC2, SC1, and SC2 cells from the staggered and layered assembly. In PC1 and SC1 cells, tropocollagen molecules are aligned in a direction parallel to the c-axis of hydroxyapatite crystals. In PC2 and SC2 cells, tropocollagen molecules are aligned in a direction normal to the longitudinal c-axis of hydroxyapatite super cell. Dimensions of hydroxyapatite crystals are approximately the same in all cells. Tropocollagen molecules are all shown in multicolor segments, and water molecules are shown in *cyan*

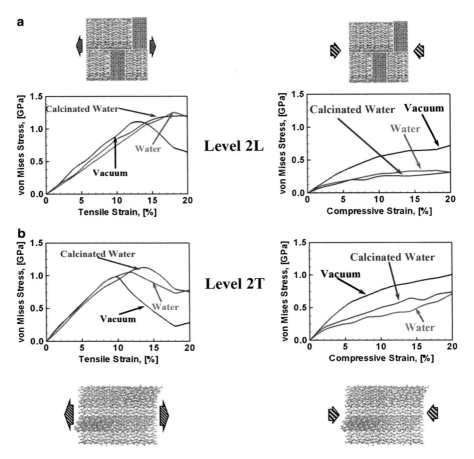

Fig. 4.5 Tensile and compressive von Mises stress vs. strain plots as a function of chemical environment in the case of (**a**) supercell Layer 2L loaded in longitudinal direction and (**b**) supercell Level 2T loaded in transverse direction

onto the crystal surface. A three-dimensional atomistic modeling framework, using NAMD molecular dynamics package [55], is developed which combines both organic and inorganic cells together to form a supercell as shown in Fig. 4.4. Two levels of hierarchical simulation cells corresponding to Level n and Level $n+1$ (Fig. 4.3) are generated with two different possible types of interfaces (Fig. 4.1). Secondary supercells are formed by embedding corresponding primary supercells in TC matrix. Two different loadings, transverse and longitudinal, were applied to all supercells. For this purpose, quasi-static mechanical deformation of each supercell is performed and characteristic stress–strain curves were obtained (Fig. 4.5) [20, 21]. Also, both tensile and compressive loading cases were considered and compared (Fig. 4.5) [22]. Furthermore, the effect of hydration on such TC–HAP biomaterials was also studied. For failure analysis, TSC model [25, 33] was used to estimate the interfacial shear strength and fracture localization zone width.

Using a MD simulation framework, Young's modulus value of TC molecule was obtained to be ~9 GPa, which is a fairly good estimate and lies in the range of modulus values obtained by other researchers [20]. The analyses confirm that the relative alignment of TC molecules with respect to the HAP mineral surface such that the interfacial contact area is maximized and the optimal direction of applied loading with respect to the TC–HAP interface orientation are important factors that contribute to making nanoscale staggered arrangement a preferred structural configuration in such biological materials. The analyses also point out that such an arrangement results in higher interfacial strength as well as higher fracture strength. In addition, such TC–HAP nanocomposite shows toughening and strain-hardening behavior, which is attributed to the reconstitution of columbic interactions between TC and HAP at the interface during sliding. The dominant tensile failure mechanism at the HAP–TC interface is simply the interfacial separation of TC and HAP without significant initial HAP deformation (Fig. 4.6). The NH_3^+ and COO^- groups in TC molecules are strongly attracted to the ions in HAP surface

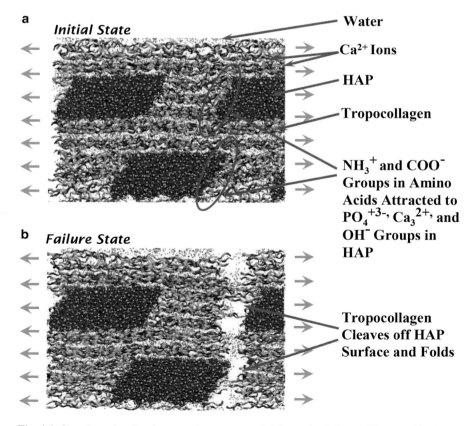

Fig. 4.6 Snapshots showing the two primary stages of deformation in Level 2T supercell. Almost invariably the peak of the stress–strain curves coincides with the collagen cleaving from the hydroxyapatite–tropocollagen interface. Similar, cleavage-driven interfacial separation mechanism was observed for Level 1L, Level 1T, and Level 2L as well

(Ca^{2+}, PO_4^{3-}, and OH^- ions) [56]. Since TC is a flexible chain-like molecule, it elongates on applied deformation but cleaves off after the point when it is fully stretched. Such cleavage results in local nanoscale interfacial failure.

Most biological materials in physiological systems contain water as one of their constituents. It has been observed that the presence of water molecules in the TC–HAP biocomposites enhanced overall composite mechanical strength (Fig. 4.5a) [20]. This is attributed to water molecule's affinity for charged surfaces, such as HAP surface [56], and NH_3^+ and COO^- groups in TC, owing to its polar nature and capability of make strong hydrogen bonds. As a result, water acts as an electrostatic bridge between HAP surface and TC molecules and strengthens the TC–HAP interface. This strengthening especially plays an important role whenever there is a relative sliding occurs between HAP surface and TC molecules at the interface. Previous studies have shown that hydration has a stabilizing effect on the collagen triple helix [57], and solvated TC molecule requires more energy to untie from the HAP surface [58]. Similar interaction behavior of water at protein–mineral interface is found in nacre as well [59, 60]. It also acts as glue between TC–TC interactions [47] and, thereby, delays the failure of the overall system.

Another interesting finding has emerged out of a comparison between stress–strain curves for two different hierarchical levels, primary cells (PS cells, level I) and secondary cells (SC cells, level II) (Figs. 4.4 and 4.5). The ultimate strength values for primary cells are an order higher than the values for secondary cells. However, the strain values corresponding to peak stress for both PS and SC cells are of the same order and in the same range. This suggests that failure of such biocomposites is predominantly strain dependent and not a function of ultimate strength. In a fractal bone model [43] (Fig. 4.3), scaling relations are developed for obtaining both ultimate strength values and Young's modulus values between two hierarchical levels. The strength and stiffness values obtained using MD-based analyses are in good agreement with these scaling relations [22].

The MD framework of Dubey and Tomar is quite versatile and is capable of analyzing any organic–inorganic nanostructure with different geometry, polypeptide sequence, and organization for its mechanical deformation and failure behavior. There is some argument as to whether the crystalline mineral in bone is needle shaped or plate shaped [35, 61–63]. Hence, Dubey and Tomar have investigated the effect of plate- and needle-shaped HAP crystals on the mechanical strength of TC–HAP biomaterials. Plate-shaped cells (PS) and needle-shaped cells (NS) were generated (Fig. 4.8), and stress–strain curves were obtained for longitudinal and transverse loading (Fig. 4.9). NS cells showed higher ultimate strength values as compared to PS cells, but overall less toughening and strain-hardening behavior. PS cells also showed higher interfacial strength and fracture resistance as compared to NS cells [53, 64]. Visual deformation analysis using VMD (visual molecular dynamics) package revealed another interesting observation: HAP mineral platelets deform in a ductile fashion during compressive deformation of TC–HAP biocomposite (Fig. 4.7). This is attributed to the fact that the HAP crystal size here is very

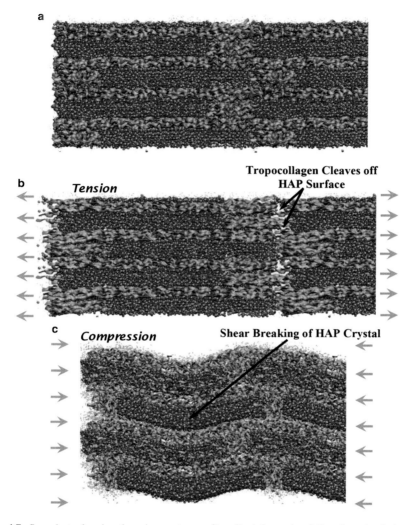

Fig. 4.7 Snapshots showing the primary stages of tensile deformation failure in a simulation cell with plate-shaped HAP crystals with loading in longitudinal direction (PS-L cell), with (**a**) cell at 0 % strain value, and (**b**) cell at 14 % strain value. Almost invariably the peak of the stress–strain curves coincides with the TC molecules cleaving off from the HAP–TC interface. Similar, cleavage-driven interfacial separation mechanism was observed for simulation cell with needle-shaped HAP crystal (NS-L cell) tensile deformation as well. (**c**) PS-L cell under compressive deformation. The HAP crystals deform in a highly ductile fashion. This can be attributed to the fact that the HAP crystal size here is very small and thickness is only few monolayers. Therefore, unlike a bigger bulk of HAP material, where external electrostatic force penetration is confined only to the bulk surface, the electrostatic forces in this case have influenced the whole HAP crystal leading to such behavior

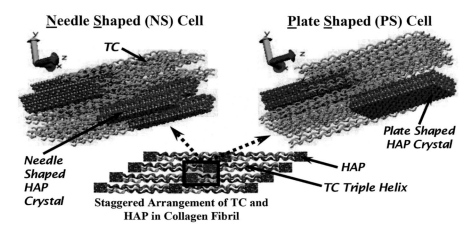

Needle Shaped (NS) Cell   **Plate Shaped (PS) Cell**

TC

*Plate Shaped
HAP Crystal*

*Needle
Shaped
HAP
Crystal*   **Staggered Arrangement of TC and
HAP in Collagen Fibril**

HAP

TC Triple Helix

Fig. 4.8 A schematic showing the derivation of two simulation cells, one with HAP crystals in *needle* shape (NS) and other with HAP crystals in *plate* shape (PS), from the staggered and layered TC–HAP assembly in center. Tropocollagen molecules are shown in *green wavy ribbons*. The loading direction is along the "*Z*"-axis of the simulation cells, i.e., parallel to TC longitudinal axes. In both cells, the *c*-axes (0 0 1) of HAP crystals are aligned parallel to "*Z*"-axis

small and thickness is only few monolayers. Therefore, unlike a bigger bulk of HAP material, where external electrostatic force penetration is confined only to the bulk surface, the electrostatic forces in this case have influenced the whole HAP crystal leading to such behavior.

A further analysis using the MD model has been performed to gain valuable insight into correlation between structure–property relations and HAP mineral distribution and nucleation, in the context of bone disease such as osteogenesis imperfecta (OI) mutations [54]. TC–HAP biomaterial is analyzed for its mechanical strength and failure behavior with eight different mutated TC sequences, out of which five of them correspond to lethal OI mutations [65, 66]. Two different mineral distributions, HAP crystals with needle shape (NS) and with plate shape (PS) (Fig. 4.8), were considered for all mutations. Results showed that the effect of OI mutations on the strength of TC–HAP biomaterials is insignificant (Fig. 4.10). This analysis points out that the HAP mineral morphology and distribution plays a stronger role than the stiffness loss of single mutated TC molecule, in affecting the strength, toughness, and stiffness values of model TC–HAP biocomposites at the nanoscopic length scale. This leads us to think that if mutations in TC are not directly affecting the TC–HAP biocomposite mechanical strength significantly, then they are probably playing a role indirectly by changing the HAP mineral morphology and distribution of OI bone in its growth and nucleation period.

It is clear that interfaces play an important role not only in biocomposites but in synthetic composites as well. More often the constituents of such composites are formed and organized at nanoscopic length scales, and it becomes increasingly difficult to study the interfacial properties by experiments. In such scenarios, computational approaches such as molecular dynamics simulations prove to be powerful tool of investigation. However, MD also has limitations in many aspects.

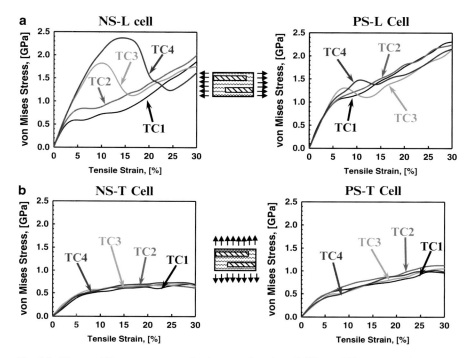

Fig. 4.9 The von Mises stress vs. strain plots as a function of different TC sequences in the case of (**a**) cells NS-L and PS-L and (**b**) cells NS-T and PS-T

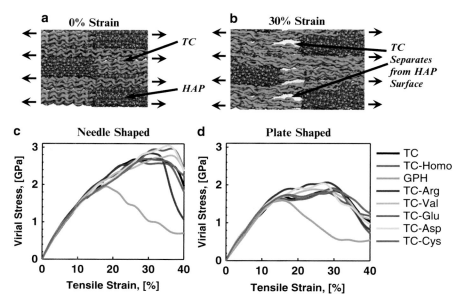

Fig. 4.10 (**a**, **b**) Close-up snapshots showing the primary stages of tensile deformation failure in NS cell (having TC molecules), with cell at 0 % strain value and cell at 30 % strain value. Almost invariably the peak of the stress–strain curves coincides with the tropocollagen molecules separating from the hydroxyapatite–tropocollagen interface. (**c**, **d**) Virial stress vs. tensile strain curve for (**c**) NS cells and (**d**) PS cells, as a function of different TC polypeptides. Strength is substantially diminished for Gly-Pro-Hyp triplet homotrimer (*green curve*)

MD uses empirical potentials for force calculations and therefore approximates the dynamics between two atoms such as bond breaking and forming. In order to be able to resolve the atomic vibrations, the time steps are of the order of femtoseconds, which limits the system size and total simulation time to nanoseconds. For example, in the abovementioned study, a segment of full-length TC molecule is used, which is otherwise a very big molecule (~300 nm) and infeasible to simulate. These limitations, though unavoidable due to the intrinsic nature of the molecular dynamics methodology and the available computational capacity, should be kept in mind. The experimental techniques closest to measurement and visualization of such small-scale interfacial properties are atomic force microscopy (AFM) and SEM. However, (1) instrument sensitivity and resolution and (2) sample integrity after preparation and during testing are two major roadblocks in such techniques. Small-angle X-ray scattering-based studies have been useful in structural characterization. But the use of such techniques is very limited for mechanical strength characterization.

4.2.2   Nacre

Nacre is found in the interior layers of the hard outer shells of mollusks. The exterior layers of the shell are typically brittle, but hard, and provide resistance to penetration from external impact [67, 68], while nacre, present in the inner layers of the shell, provides toughening by dissipating the mechanical energy owing to its capability of undergoing large inelastic deformations [69]. Main constituents of nacre are aragonite (a crystallographic form of calcium carbonate, 95 % by volume) and elastic organic material (proteins such as chitin and polysaccharides). Nacre has a hierarchical brick-and-mortar structure (Fig. 4.11) with polygonal aragonite tablets as bricks and organic matrix as the mortar acting as the filler and adhesive between such tablets [72]. The aragonite tiles are roughly 0.5 μm thick and 5–8 μm in diameter. In some of the cases, like red abalone, nacre shows some level of organization across different layers, where tablets are stacked in columns with some overlap between tablets from adjacent columns (Fig. 4.11d). The interface between the tablets is roughly 30 nm thick and is composed of organic materials [73], nanoasperities [69], and direct mineral bridges connecting two adjacent tablets [74, 75]. In fact, the tablets themselves are formed by aragonite nanograins (Fig. 4.11g), which again are separated by a very fine three-dimensional network of organic material [76, 77].

Aragonite tablets and organic material in nacre are organized in such a fashion that if a load is applied perpendicular to the plane of tablets, the organic material arrests the uncontrolled growth of crack by deforming inelastically and dissipating energy [78–80]. In this mechanism, the tablets approximately remain linear elastic, and the large deformations are provided by significant shearing of the organic–inorganic interface. The presence of organic material between nanograins in a single tablet might also provide some resistance to tablet sliding. However, this viscoelastic mechanism is active only in hydrated state of nacre. A comparison of measurement

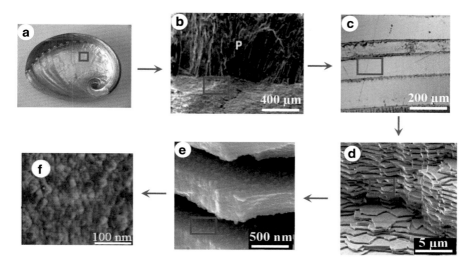

Fig. 4.11 Structural hierarchy in nacre, showing at least six structural levels. Some images were borrowed with permission from (**b, c, f**) [70], (**d**) [71], and (**e**) [16]

of mechanical properties of nacre [3, 69, 70, 78, 81, 82] and mechanical properties of individual tablets of nacre using nanoindentation [76, 82, 83] has shown that both of them have properties similar to single crystal aragonite, which is stiff but brittle. Considering that 95 % of nacre is aragonite, dry nacre behaves very much similar to pure aragonite because organic material becomes brittle. Apart from the above mechanism, few other mechanisms and factors are also proposed, which contribute to hardening, energy dissipation, and damage distribution. Similar to protein unfolding domains in bone [28, 84], it is suggested that proteins in nacre also have folded modules and act as high-performance adhesives, binding the tablets and possibly the nanograins together [79]. During deformation under loading, these biopolymers elongate in stepwise fashion and absorbing significant amount of energy. The interlocking between mineral platelets [85] increases toughness by progressively failing and limiting the catastrophic failure [86]. The waviness found in mineral tablets can be significant [17] and progressively locks tablet sliding, leading to a more difficult separation of tablets from their interfaces.

Efforts for developing composites which can emulate nacre have been in the forefront of hard biomimetic materials development. Kotov and coworkers [87] developed a method using polyelectrolytes and clays to make a nanoscale version of nacre. They used the approach of layer-by-layer assembly (LBL) of externally controlled sequential deposition of organic and inorganic layers and resulted in a nanocomposite material which resembled the brick-and-mortar arrangement similar to nacre along with the crucial effect of the sacrificial bonds (Fig. 4.12). This composite's ultimate tensile strength and Young's modulus values approached that of nacre and lamellar bones, respectively. This idea opened the door for development of similar nacre-like bio-inspired materials, such as bone implants. LBL assembly

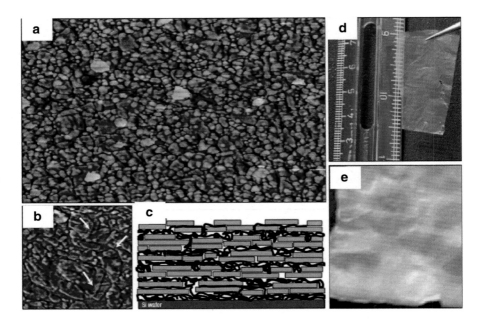

Fig. 4.12 Microscopic and macroscopic description of polyelectrolyte–clay multilayers ($(P/C)_n$). (**a**) Phase contrast AFM image of a $(P/C)_1$ film on silicon substrate. (**b**) Enlarged portion of the film in a showing overlapping clay platelets marked by arrows. (**c**) The $(P/C)_n$ film structure. The thickness of each clay platelet is 0.9 nm. (**d**) Photograph of free-standing $(P/C)_{50}$ film after delamination. (**e**) Close-up photograph of the film in (**d**) under side illumination. Image is borrowed with permission from [87]

method has the ability to incorporate multifunctionality in the material owing to the robust control over deposition of every layer. With this idea, a similar hard nano-biocomposite material was developed with antibacterial activity [88]. The material was structurally stable, inhibited growth of *E. coli* for over 18 h, and did not prevent growth of the mammalian tissue.

The above method is good for achieving tablet thicknesses of the order of the aragonite tablet thickness in nacre (~0.5 μm). However, such methods are limited to fabrication of only thin films. On the other hand, if conventional methods are used for bulk material fabrication, they end up generating mineral tablets more than two orders of magnitude larger than 0.5 μm [89]. In a recent effort to overcoming both the above hurdles, Ritchie and coworkers [16, 18] fabricated bulk nacre-like composite material using ice-templated structures, which resulted in tablet thickness of ~5 μm, Fig. 4.13a–d. The base materials used here are alumina (Al_2O_3) and PMMA (polymethyl methacrylate). The toughness of such composite can be at least two orders of magnitude higher than its constituents, which is primarily attributed to the interfacial design and chemo-mechanics of the interface. It has properties comparable to those of aluminum alloys. A comparison between the toughening mechanisms acting in artificial and natural nacre is shown in Fig. 4.13. By adding chemicals

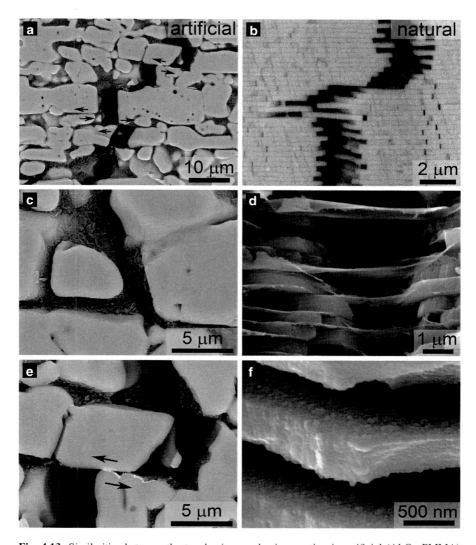

Fig. 4.13 Similarities between the toughening mechanisms acting in artificial (Al_2O_3–PMMA) and natural nacre. Scanning electron micrographs taken during an in situ R-curve measurement of (**a**, **c**, and **e**) a grafted brick-and-mortar structure and (**b**, **d**, and **f**) of hydrated nacre (abalone shell) show similar mechanisms although the nacre clearly has a finer structure. The image taken during stable crack propagation shows the toughening mechanisms acting at multiple length scales: (**a**) "pullout" mechanisms similarly to that observed in (**b**) nacre, (**c**) polymer tearing and stretching over micrometer dimensions as also observed in the (**d**) organic phase of nacre, and (**e**) frictional sliding resisted by the interface roughness of the ceramic bricks as observed in (**f**) nacre. Note that the thin bright lines between the sliding grains in (**e**) result from the deformation of the gold coating during sliding; indeed, all the mechanisms described above are rendered visible by electrical charging in the SEM resulting from the deformation of the gold coating during deformation. Image is borrowed with permission from [16]

such as sucrose to the above fabrication method, it was possible to make mineral bridges similar to mineral connections found in nacre. Such model materials and techniques provide potential platform and possibilities for development of the science fabrication of new age materials of this twenty-first century.

4.3   Discussion: Development of Materials with Controlled Nanoscale Interfacial Design

A fundamental milestone in development of biomimetic composite materials is finding a synthetic pathway to artificial analogs of hard biological materials. Efforts for developing composites which can emulate biological materials have gained high momentum during the past decade. The materials of nacre and bones are well known for their hardness, strength, multifunctionality, and toughness and are superior to many man-made ceramics and composites. The structure–function compatibility of such materials has inspired a big class of biomimetic advanced materials and organic–inorganic composites. The main cause for the high toughening properties of nacre has been claimed to be the consequence of the highly tortuous fracture path occurring in the material with the exposing of the surface of ceramic platelets. However, the high amount of energy dissipation experimentally measured cannot be explained by just considering the simple brick-and-mortar architecture of nacre.

An increase in the ratio of interfacial area to system volume in materials developed based on hard biomaterial architecture can lead either ways, weakening or strengthening, due to the lack of a systematic understanding in overall material organization. It is beyond the current state of the art in materials technology to make a hierarchical biocomposite material using the same constituents as used by the nature. Inspired from the self-healing capability of bone, recently self-healing polymers have been developed [90]. These polymers contain pockets of a healing agent and a catalyst. A crack propagating through the material will puncture these pockets and release, mix, and cure these agents, effectively healing the crack by filling it with glue. The original load-carrying capability of the material could be restored and, in some cases, even improved. While this mechanism is not as sophisticated as bone, it is a promising example of a self-healing artificial material.

4.3.1   Biomaterials Inspired from Interfacial Design of Nacre and Bone

The ordered brick-and-mortar arrangement is considered to be the key strength and toughness determining structural feature of nacre. It has also been shown that the ionic cross-linking of tightly folded macromolecules is equally important. Tang et al. [87] have demonstrated that both structural features can be reproduced by sequential deposition of polyelectrolytes and clays. The resulting organic–inorganic hybrid reproduced both the brick-and-mortar arrangement and the crucial effect of the sacrificial bonds. The ultimate tensile strength and Young's modulus of such a composite approached that of nacre and lamellar bones, respectively. This makes such materials potential candidates for bone implant applications. Further, this layer-by-layer assembly is used to develop antibacterial biocompatible version of the above organic–inorganic material [88].

The architecture of nacre is such that if a stress normal to the platelet plane is applied, then the ductile organic fraction that glues the crystals will prevent an uncontrolled crack growth [80]. This viscoelastic dissipative process takes place mainly in the hydrated state, and dry nacre behaves very much similar to pure aragonite because the organic material becomes brittle [17]. The transition from a brittle (glassy) state to a more elastomeric behavior was observed in a polysaccharide (chitosan) upon hydration using nonconventional dynamic mechanical analysis, in which a glass transition could be detected in intermediate hydration levels [91]. Photoacoustic Fourier transform infrared spectroscopy results suggest that the water present at the nanograin interfaces also contributes significantly to the viscoelasticity of nacre [92]. It was also suggested that proteins composing the organic matrix, such as Lustrin A, exhibit a highly modular structure characterized by a multidomain architecture with folded modules and act as a high-performance adhesive, binding the platelets together [93]. Upon increasing stress, these biopolymers elongate in a stepwise manner as folded domains or loops are pulled open, with significant energy requirements to unfold each individual module—it was claimed that this modular elongation mechanism may contribute to the amazing toughness properties in nacre [93]. Other toughening pathways are also present such as the presence of asperities onto the surface of the aragonite tablets [94], mineral bridges [74], interlocks [95], waved surfaces [17], nanoscale nature and organization of the building blocks constituting nacre [7], and rotation and deformation of aragonite nanograins [96].

The overall function and properties of hard biological composite materials are a complicated coupled function of the structural arrangement at different hierarchical levels and individual constituent material properties. In general, it is impractical to separate structure and properties of such biocomposites to explain the material behavior. The success in mimicking even relatively simple structures, such as nacre, has been relatively modest yet. The best examples [87, 88] are far away from providing a millimeter-thick artificial nacre layer with greatly enhanced fracture toughness, a relatively simple task a mollusk has been routinely doing during its life cycle. Unlike artificial nanocomposite materials built using optimal interfacial arrangements, biological materials show remarkable strength compared to their relatively weaker constituents. In fact, sometimes artificial nanocomposites can turn out to be weaker than its constituents. Two important aspects of the biological material design arise here, which differentiate the artificial and natural nanocomposites: (1) prevalence of carefully designed organic–inorganic interfaces at almost all hierarchical levels and (2) multilevel hierarchy with different microstructure designs at different levels. The first aspect points out that biological materials have extremely large amount of interfacial area. Moreover, these interfaces are formed by specific relative arrangement between the organic phases and inorganic phases for optimum load handling. Second, different microstructural designs at each level provide further optimization/customization parameters for desired mechanical performance. This suggests that an in-depth understanding of chemo-mechanics of such interfaces, their role in load handling, and failure mechanisms at the fundamental length scales is vital to the development of both the bioengineering and biomimetics fields.

Acknowledgments The authors acknowledge the partial support from the National Science Foundation and express sincere gratitude to authors and publishers of the papers whose figures are cited in the manuscript.

References

1. K. Autumn et al., Adhesive force of a single gecko foot-hair. Nature **405**, 681–685 (2000)
2. H.J. Gao, X. Wang, H.M. Yao, S. Gorb, E. Arzt, Mechanics of hierarchical adhesion structures of geckos. Mech. Mater. **37**, 275–285 (2005). doi:10.1016/j.mechmat.2004.03.008
3. J.D. Currey, Mechanical-properties of mother of pearl in tension. Proc. R. Soc. Lond. B Biol. Sci. **196**, 443–463 (1977)
4. J.L. Katz, Hierarchical Modeling of Compact Haversian Bone as a Fiber Reinforced Material, in *Advances in Bioengineering*, ed. by R.E. Mates, C.R. Smith, vol. 1976 (ASME, New York, NY, 1976), pp. 17–18. *Meeting*, New York, NY, USA, 5–10 Dec 1976. Vi+42p. Illus
5. J.L. Katz, Anisotropy of Young's modulus of bone. Nature **283**, 106–107 (1980)
6. H.J. Gao, B.H. Ji, I.L. Jager, E. Arzt, P. Fratzl, Materials become insensitive to flaws at nanoscale: lessons from nature. Proc. Natl. Acad. Sci. U. S. A. **100**, 5597–5600 (2003). doi:10.1073/pnas.0631609100
7. P. Fratzl, R. Weinkamer, Nature's hierarchical materials. Prog. Mater. Sci. **52**, 1263–1334 (2007)
8. M.A. Meyers, P.Y. Chen, A.Y.M. Lin, Y. Seki, Biological materials: structure and mechanical properties. Prog. Mater. Sci. **53**, 1–206 (2008). doi:10.1016/j.pmatsci.2007.05.002
9. J.Y. Rho, L. Kuhn-Spearing, P. Zioupos, Mechanical properties and the hierarchical structure of bone. Med. Eng. Phys. **20**, 92–102 (1998)
10. M.E. Launey, R.O. Ritchie, On the fracture toughness of advanced materials. Adv. Mater. **21**, 2103–2110 (2009). doi:10.1002/adma.200803322
11. M. Sarikaya, C. Tamerler, A.K.Y. Jen, K. Schulten, F. Baneyx, Molecular biomimetics: nanotechnology through biology. Nat. Mater. **2**, 577–585 (2003)
12. B.D. Ratner, S.J. Bryant, Biomaterials: where we have been and where we are going. Annu. Rev. Biomed. Eng. **6**, 41–75 (2004). doi:10.1146/annurev.bioeng.6.040803.140027
13. C. Sanchez, H. Arribart, M.M.G. Guille, Biomimetism and bioinspiration as tools for the design of innovative materials and systems. Nat. Mater. **4**, 277–288 (2005). doi:10.1038/nmat1339
14. P. Fratzl, Biomimetic materials research: what can we really learn from nature's structural materials? J. R. Soc. Interface **4**, 637–642 (2007). doi:10.1098/rsif.2007.0218
15. B. Bhushan, Biomimetics: lessons from nature—an overview. Philos. Trans. R. Soc. A Math. Phys. Eng. Sci. **367**, 1445–1486 (2009). doi:10.1098/rsta.2009.0011
16. M.E. Launey et al., Designing highly toughened hybrid composites through nature-inspired hierarchical complexity. Acta Mater. **57**, 2919–2932 (2009). doi:10.1016/j.actamat.2009.03.003
17. F. Barthelat, Biomimetics for next generation materials. Philos. Trans. R. Soc. A Math. Phys. Eng. Sci. **365**, 2907–2919 (2007). doi:10.1098/rsta.2007.0006
18. E. Munch et al., Tough, bio-inspired hybrid materials. Science **322**, 1516–1520 (2008). doi:10.1126/science.1164865

19. L.C. Palmer, C.J. Newcomb, S.R. Kaltz, E.D. Spoerke, S.I. Stupp, Biomimetic systems for hydroxyapatite mineralization inspired by bone and enamel. Chem. Rev. **108**, 4754–4783 (2008). doi:10.1021/cr8004422

20. D.K. Dubey, V. Tomar, Role of the nanoscale interfacial arrangement in mechanical strength of tropocollagen-hydroxyapatite based hard biomaterials. Acta Biomater. **5**, 2704–2716 (2009). doi:10.1016/j.actbio.2009.02.035

21. D.K. Dubey, V. Tomar, Understanding the influence of structural hierarchy and its coupling with chemical environment on the strength of idealized tropocollagen–hydroxyapatite biomaterials. J. Mech. Phys. Solid **57**, 1702–1717 (2009). doi:10.1016/j.jmps.2009.07.002

22. D.K. Dubey, V. Tomar, The effect of tensile and compressive loading on the hierarchical strength of idealized tropocollagen-hydroxyapatite biomaterials as a function of the chemical environment. J. Phys. Condens. Matter **21**, 205103 (2009). doi:10.1088/0953-8984/21/20/205103

23. T. Leventouri, Synthetic and biological hydroxyapatites: crystal structure questions. Biomaterials **27**, 3339–3342 (2006)

24. S.C. Cowin, *Bone Mechanics Handbook* (CRC Press, Boca Raton, FL, 2001)

25. B. Ji, H. Gao, Mechanical properties of nanostructure of biological materials. J. Mech. Phys. Solid **52**, 1963–2000 (2004)

26. S. Weiner, H.D. Wagner, The material bone: structure mechanical function relations. Annu. Rev. Mater. Sci. **28**, 271–298 (1998)

27. P.J. Thurner et al., High-speed photography of the development of microdamage in trabecular bone during compression. J. Mater. Res. **21**, 1093–1100 (2006). doi:10.1557/jmr.2006.0139

28. G.E. Fantner et al., Sacrificial bonds and hidden length dissipate energy as mineralized fibrils separate during bone fracture. Nat. Mater. **4**, 612–616 (2005)

29. I. Jager, P. Fratzl, Mineralized collagen fibrils: a mechanical model with a staggered arrangement of mineral particles. Biophys. J. **79**, 1737–1746 (2000)

30. P. Fratzl, H.S. Gupta, E.P. Paschalis, P. Roschger, Structure and mechanical quality of the collagen-mineral nano-composite in bone. J. Mater. Chem. **14**, 2115–2123 (2004)

31. H.S. Gupta et al., Cooperative deformation of mineral and collagen in bone at the nanoscale. Proc. Natl. Acad. Sci. U. S. A. **103**, 17741–17746 (2006). doi:10.1073/pnas.0604237103

32. H.S. Gupta et al., Nanoscale deformation mechanisms in bone. Nano Lett. **5**, 2108–2111 (2005)

33. B.H. Ji, A study of the interface strength between protein and mineral in biological materials. J. Biomech. **41**, 259–266 (2008)

34. P. Fratzl, N. Fratzlzelman, K. Klaushofer, G. Vogl, K. Koller, Nucleation and growth of mineral crystals in bone studied by small-angle X-ray scattering. Calcif. Tissue Int. **48**, 407–413 (1991)

35. W.J. Landis, K.J. Hodgens, J. Arena, M.J. Song, B.F. McEwen, Structural relations between collagen and mineral in bone as determined by high voltage electron microscopic tomography. Microsc. Res. Tech. **33**, 192–202 (1996)

36. H.R. Wenk, F. Heidelbach, Crystal alignment of carbonated apatite in bone and calcified tendon: results from quantitative texture analysis. Bone **24**, 361–369 (1999)

37. S.J. Eppell, B.N. Smith, H. Kahn, R. Ballarini, Nano measurements with micro-devices: mechanical properties of hydrated collagen fibrils. J. R. Soc. Interface **3**, 117–121 (2005)

38. H.S. Gupta et al., Fibrillar level fracture in bone beyond the yield point. Int. J. Fract. **139**, 425–436 (2006)

39. N. Sasaki, S. Odajima, Elongation mechanism of collagen fibrils and force-strain relations of tendon at each level of structural hierarchy. J. Biomech. **29**, 1131–1136 (1996)

40. N. Sasaki, S. Odajima, Stress-strain curve and Young's modulus of a collagen molecule as determined by the X-ray diffraction technique. J. Biomech. **29**, 655–658 (1996)

41. A.J. Hodge, J.A. Petruska, in *Aspects of Protein Structure. Proceedings of a Symposium*, ed. by G.N. Ramachandran (Academic Press, Inc., London, New York, 1963), pp. 289–300

42. P.J. Thurner et al., High-speed photography of compressed human trabecular bone correlates whitening to microscopic damage. Eng. Fract. Mech. **74**, 1928–1941 (2007)

43. H. Gao, Application of fracture mechanics concepts to hierarchical biomechanics of bone and bone-like materials. Int. J. Fract. **138**, 101–137 (2006)
44. A.C. Lorenzo, E.R. Caffarena, Elastic properties, Young's modulus determination and structural stability of the tropocollagen molecule: a computational study by steered molecular dynamics. J. Biomech. **38**, 1527–1533 (2005)
45. M. Israelowitz, S.W.H. Rizvi, J. Kramer, H.P. von Rizvi, Computational modeling of type I collagen fibers to determine the extracellular matrix structure of connective tissues. Protein Eng. Des. Sel. **18**, 329–335 (2005)
46. J.W. Handgraaf, F. Zerbetto, Molecular dynamics study of onset of water gelation around the collagen triple helix. Proteins Struct. Funct. Bioinform. **64**, 711–718 (2006)
47. D. Zhang, U. Chippada, K. Jordan, Effect of the structural water on the mechanical properties of collagen-like microfibrils: a molecular dynamics study. Ann. Biomed. Eng. **35**, 1216–1230 (2007)
48. R.J. Radmer, T.E. Klein, Triple helical structure and stabilization of collagen-like molecules with 4(R)-hydroxyproline in the Xaa position. Biophys. J. **90**, 578–588 (2006)
49. T. Hassenkam et al., High-resolution AFM imaging of intact and fractured trabecular bone. Bone **35**, 4–10 (2004)
50. W.J. Landis et al., Mineralization of collagen may occur on fibril surfaces: evidence from conventional and high-voltage electron microscopy and three-dimensional imaging. J. Struct. Biol. **117**, 24–35 (1996)
51. S. Weiner, Y. Talmon, W. Traub, Electron diffraction of mollusc shell organic matrices and their relationship to the mineral phase. Int. J. Biol. Macromol. **5**, 325–328 (1983)
52. W.J. Landis, M.J. Song, A. Leith, L. McEwen, B.F. McEwen, Mineral and organic matrix interaction in normally calcifying tendon visualized in 3 dimensions by high-voltage electron-microscopic tomography and graphic image-reconstruction. J. Struct. Biol. **110**, 39–54 (1993)
53. D.K. Dubey, V. Tomar, Role of hydroxyapatite crystal shape in nanoscale mechanical behavior of model tropocollagen-hydroxyapatite hard biomaterials. Mater. Sci. Eng. C Mater. Biol. Appl. **29**, 2133–2140 (2009). doi:10.1016/j.msec.2009.04.015
54. D.K. Dubey, V. Tomar, Effect of osteogenesis imperfecta mutations in tropocollagen molecule on strength of biomimetic tropocollagen-hydroxyapatite nanocomposites. Appl. Phys. Lett. **96**, 023703 (2010)
55. J.C. Phillips et al., Scalable molecular dynamics with NAMD. J. Comput. Chem. **26**, 1781–1802 (2005)
56. A.S. Posner, R.A. Beebe, The surface chemistry of bone mineral and related calcium phosphates. Semin. Arthritis Rheum. **4**, 267–291 (1975)
57. A.D. Simone, L. Vitaglaino, R. Berisio, Role of hydration in collagen triple helix stabilization. Biochem. Biophys. Res. Commun. **372**, 121–125 (2008)
58. R. Bhowmik, K.S. Katti, D.R. Katti, Influence of mineral-polymer interactions on molecular mechanics of polymer in composite bone biomaterials. Mater. Res. Soc. Symp. Proc. **978**, 6 (2007)
59. F. Barthelat, H.D. Espinosa, An experimental investigation of deformation and fracture of nacre-mother of pearl. Exp. Mech. **47**, 311–324 (2007). doi:10.1007/s11340-007-9040-1
60. P. Ghosh, D.R. Katti, K.S. Katti, Mineral proximity influences mechanical response of proteins in biological mineral-protein hybrid systems. Biomacromolecules **8**, 851–856 (2007)
61. E. Bonucci, *Mechanical Testing of Bone and the Bone–Implant Interface* (CRC Press, Boca Raton, FL, 2000)
62. J.D. Currey, *Bones: Structure and Mechanics* (Princeton University Press, Princeton, 2002)
63. N. Matsushima, M. Akiyama, Y. Terayama, Quantitative-analysis of the orientation of mineral in bone from small-angle X-ray-scattering patterns. Jpn. J. Appl. Phys. **21**, 186–189 (1982)
64. D.K. Dubey, V. Tomar, Effect of changes in tropocollagen residue sequence and hydroxyapatite mineral texture on the strength of ideal nanoscale tropocollagen-hydroxyapatite biomaterials. J. Mater. Sci. Mater. Med. **21**, 161–171 (2010)

65. A. Gautieri, S. Vesentini, A. Redaelli, M.J. Buehler, Single molecule effects of osteogenesis imperfecta mutations in tropocollagen protein domains. Protein Sci. **18**, 161–168 (2009). doi:10.1002/pro.21

66. D.L. Bodian, B. Madhan, B. Brodsky, T.E. Klein, Predicting the clinical lethality of osteogenesis imperfecta from collagen glycine mutations. Biochemistry **47**, 5424–5432 (2008). doi:10.1021/bi800026k

67. J.D. Currey, J.D. Taylor, The mechanical behaviour of some molluscan hard tissues. J. Zool. **173**, 395–406 (1974)

68. M. Sarikaya, I.A. Aksay, *Biomimetic, Design and Processing of Materials Polymers and Complex Materials* (American Institute of Physics, Woodbury, NY, 1995)

69. R.Z. Wang, Z. Suo, A.G. Evans, N. Yao, I.A. Aksay, Deformation mechanisms in nacre. J. Mater. Res. **16**, 2485–2493 (2001)

70. R. Menig, M.H. Meyers, M.A. Meyers, K.S. Vecchio, Quasi-static and dynamic mechanical response of Haliotis rufescens (abalone) shells. Acta Mater. **48**, 2383–2398 (2000)

71. N. Yao, A. Epstein, A. Akey, Crystal growth via spiral motion in abalone shell nacre. J. Mater. Res. **21**, 1939–1946 (2006). doi:10.1557/jmr.2006.0252

72. M. Sarikaya, An introduction to biomimetics—a structural viewpoint. Microsc. Res. Tech. **27**, 360–375 (1994)

73. T.E. Schaffer et al., Does abalone nacre form by heteroepitaxial nucleation or by growth through mineral bridges? Chem. Mater. **9**, 1731–1740 (1997)

74. F. Song, X.H. Zhang, Y.L. Bai, Microstructure and characteristics in the organic matrix layers of nacre. J. Mater. Res. **17**, 1567–1570 (2002)

75. F. Song, X.H. Zhang, Y.L. Bai, Microstructure in a biointerface. J. Mater. Sci. Lett. **21**, 639–641 (2002)

76. X.D. Li, W.C. Chang, Y.J. Chao, R.Z. Wang, M. Chang, Nanoscale structural and mechanical characterization of a natural nanocomposite material: the shell of red abalone. Nano Lett. **4**, 613–617 (2004). doi:10.1021/nl049962k

77. M. Rousseau et al., Multiscale structure of sheet nacre. Biomaterials **26**, 6254–6262 (2005). doi:10.1016/j.biomaterials.2005.03.028

78. A.P. Jackson, J.F.V. Vincent, R.M. Turner, The mechanical design of nacre. Proc. R. Soc. Lond. B Biol. Sci. **234**, 415–440 (1988)

79. B.L. Smith et al., Molecular mechanistic origin of the toughness of natural adhesives, fibres and composites. Nature **399**, 761–763 (1999)

80. T. Sumitomo, H. Kakisawa, Y. Owaki, Y. Kagawa, In situ transmission electron microscopy observation of reversible deformation in nacre organic matrix. J. Mater. Res. **23**, 1466–1471 (2008). doi:10.1557/jmr.2008.0184

81. F. Barthelat, H. Tang, P.D. Zavattieri, C.M. Li, H.D. Espinosa, On the mechanics of mother-of-pearl: a key feature in the material hierarchical structure. J. Mech. Phys. Solid **55**, 306–337 (2007). doi:10.1016/j.jmps.2006.07.007

82. F. Barthelat, C.M. Li, C. Comi, H.D. Espinosa, Mechanical properties of nacre constituents and their impact on mechanical performance. J. Mater. Res. **21**, 1977–1986 (2006). doi:10.1557/jmr.2006.0239

83. B.J.F. Bruet et al., Nanoscale morphology and indentation of individual nacre tablets from the gastropod mollusc Trochus niloticus. J. Mater. Res. **20**, 2400–2419 (2005). doi:10.1557/jmr.2005.0273

84. J.B. Thompson et al., Bone indentation recovery time correlates with bond reforming time. Nature **414**, 773–776 (2001)

85. K.S. Katti, D.R. Katti, S.M. Pradhan, A. Bhosle, Platelet interlocks are the key to toughness and strength in nacre. J. Mater. Res. **20**, 1097–1100 (2005). doi:10.1557/jmr.2005.0171

86. K.S. Katti, D.R. Katti, Why is nacre so tough and strong? Mater. Sci. Eng. C Biomim. Supramol. Syst. **26**, 1317–1324 (2006). doi:10.1016/j.msec.2005.08.013

87. Z.Y. Tang, N.A. Kotov, S. Magonov, B. Ozturk, Nanostructured artificial nacre. Nat. Mater. **2**, 413–418 (2003). doi:10.1038/nmat906

88. P. Podsiadlo et al., Layer-by-layer assembly of nacre-like nanostructured composites with antimicrobial properties. Langmuir **21**, 11915–11921 (2005). doi:10.1021/la051284+
89. H.M. Chan, Layered ceramics: processing and mechanical behavior. Annu. Rev. Mater. Sci. **27**, 249–282 (1997)
90. S.R. White et al., Autonomic healing of polymer composites. Nature **409**, 794–797 (2001)
91. J. Benesch, J. Mano, R. Reis, Proteins and their peptide motifs in acellular apatite mineralization of scaffolds for tissue engineering. Tissue Eng. Part B Rev. **14**, 433–445 (2008)
92. D. Verma, K. Katti, D. Katti, Nature of water in nacre: a 2D Fourier transform infrared spectroscopic study. Spectrochim. Acta A Mol. Biomol. Spectrosc. **67**, 784–788 (2007)
93. B.A. Wustman, J.C. Weaver, D.E. Morse, J.S. Evans, Structure–function studies of the Lustrin a polyelectrolyte domains, RKSY and D4. Connect. Tissue Res. **44**(Suppl. 1), 10–15 (2003)
94. G.M. Luz, J.F. Mano, Biomimetic design of materials and biomaterials inspired by the structure of nacre. Philos. Trans. R. Soc. A **28**(367), 1587–1605 (2009)
95. D.R. Katti, P. Ghosh, S. Schmidt, K.S. Katti, Mechanical properties of the sodium montmorillonite interlayer intercalated with amino acids. Biomacromolecules **6**, 3267–3282 (2005)
96. X. Li, Z.-H. Xu, R. Wang, In situ observation of nanograin rotation and deformation in nacre. Nano Lett. **6**, 2301–2304 (2006)

Chapter 5
Multiscaling for Molecular Models to Predict Lab Scale Sample Properties: A Review of Phenomenological Models

Abstract One of the defining features of biological materials is that they are highly hierarchical with different structures at different length scales. Often they are complex nanocomposites of soft fibrous polymeric phase and hard mineral phase. For instance, bone has up to seven levels of hierarchy and nacre shows up to six levels of hierarchal structure. In spite of complex hierarchical structures, the smallest building blocks in such biological materials are at the nanometer length scale. The extent of interfacial interaction and the interfacial arrangement are important determinants of the structure–function property relationship of biomaterials and influence the mechanical strength substantially. Challenges lie in identifying nature's mechanisms behind imparting such properties and its pathways in fabricating and optimizing these composites. The key here is the formation of large amount of precisely and carefully designed organic–inorganic interfaces and synergy of mechanisms acting over multiple scales to distribute loads and damage, dissipate energy, and resist change in properties owing to damages such as cracking. This chapter presents a brief overview of the role of interfacial structural design and interfacial forces in imparting superior mechanical performance to hard biological materials. Focus is on understanding the underlying engineering principles of nature's materials for use in biomedical engineering and biomaterial development.

Keywords Phenomenological models • Biological materials • Effect of interfaces • Hierarchical modeling • Multiscale modeling

5.1 Introduction

Biological materials have evolved over millions of years and are often found as complex composites with superior properties compared to their relatively weak original constituents. The toughness of spider silk, the strength and lightweight of bamboos, the self-healing of bone, the high toughness of nacre, and the adhesion abilities of the gecko's feet are a few of the many examples of high-performance natural materials. Hard biomaterials such as bone, nacre, and dentin have intrigued researchers for decades for their high stiffness, toughness, multifunctionality, and self-healing capabilities. For example, nacre has 3000 times more toughness compared to its mineral constituent [3]. Tooth enamel is 1000 times stiffer than its constituent protein polymer collagen [4]. The general mechanical performance of these composites is quite remarkable. In particular, they combine two properties which are usually quite contradictory but essential for the function of these materials. Bones, for example, need to be stiff to prevent bending and buckling, but they must also be tough since they should not break catastrophically even when the load exceeds the normal range. Such hard biological materials are not only lightweight but also possess high toughness and mechanical strength.

One of the defining features of such biological composites is that they are highly hierarchical with different structures at different length scales. Often they are complex nanocomposites of soft fibrous polymeric phase and hard mineral phase. For instance, bone has up to seven levels of hierarchy [5, 6] (Fig. 5.1), and nacre shows up to six levels of hierarchal structure (Fig. 5.2). Materials such as bone and nacre have such multilevel hierarchical structural design that concept of stress concentration at flaws remains invalid, leading to flaw-tolerant structure [7]. In spite of complex hierarchical structures, the smallest building blocks in such biological materials are at the nanometer length scale. For example, at the lowest level in bone, nanometer-sized crystals of carbonate apatite are embedded in the fibrous protein collagen in a well-organized staggered arrangement (Fig. 5.1f). In nanocomposite materials, the volume fraction of the protein–mineral interface can be enormous as the mineral bits have nanoscale size. For example, in a raindrop size volume of a nanocomposite, the area of interfacial region can be as large as a football field [8]. Interfaces play crucial role in regulating the overall mechanical properties of nanocomposites. In case of hard biomaterials such as bone, dentin, and nacre, they have primarily an organic phase (e.g., tropocollagen (TC) or chitin) and a mineral phase (e.g., hydroxyapatite (HAP) or aragonite) arranged in a staggered arrangement. In bone, the crystalline mineral phase is preferentially aligned along the longitudinal axis of the polypeptide molecules permitting maximum contact area in a staggered arrangement [9–12], Fig. 5.1. The extent of interfacial interaction and the interfacial arrangement are important determinants of the structure–function property relationship of biomaterials and influence the mechanical strength substantially [13–15]. Such biological materials have been reviewed in appreciable detail, in the context of their hierarchical structure, material properties, and failure mechanisms [5, 16–18].

An important aspect to focus in biomaterial engineering of hard biomaterials is the chemo-mechanics of the organic–inorganic interfaces and its correlation with overall mechanical behavior. This understanding is vital for selecting appropriate constituents, their size scales, and their relative arrangements, which in turn is governed by the functional requirements of the composite materials. For example, three-dimensional (3D) explicit atomistic failure analyses of model tropocollagen–hydroxyapatite interfacial biomaterial (similar to material found in bone tissues) performed at the nanoscopic length scale [9, 19, 20] have pointed out that maximizing the contact area between the TC and HAP phases results in higher interfacial strength as well as higher fracture strength. Analyses have also shown that high toughness and strain-hardening behavior of such biomaterial are due to reconstitution

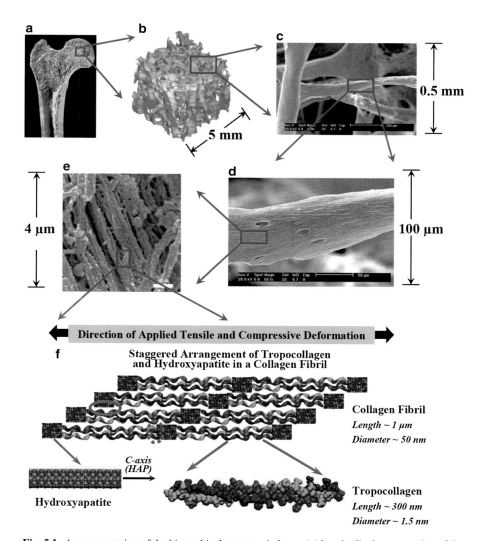

Fig. 5.1 A representation of the hierarchical structure in bone: (**a**) longitudinal cross section of the end of human femur showing the trabecular bone inside, (**b**) microstructure of trabecular bone resembling the honeycomb structure, (**c**) high-resolution view of the strut-like structures in microporous structure, (**d**) showing one such strut called trabecula, having aligned mineralized collagen fibers and lacunae (cavities left after cell apoptosis), (**e**) fracture surface of human bone showing mineralized collagen fibrils, and (**f**) a schematic of staggered and layered assembly of TC molecules and HAP platelets to form a mineralized collagen fibril. Three different colors in TC molecules depict three polypeptide chains forming a triple helix. HAP crystal's *c*-axis is along the loading direction. *Solid* and *dotted ellipses* in *pink* show two possible types of interactions between TC and HAP surface, one with TC perpendicular to HAP surface and the other with TC parallel to HAP surface. Images in (**b**), (**c**), and (**d**) are borrowed from [1] and (**e**) from [2]

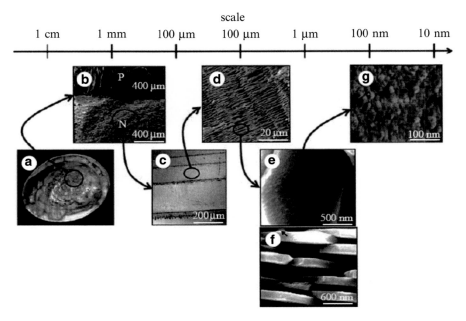

Fig. 5.2 Hierarchical organization in nacre showing at least six structural levels. This image is borrowed from [26]. However, the authors themselves borrowed parts of this image from other references

of columbic interactions between TC and HAP surfaces during interfacial sliding due to mechanical deformation. It has also been shown recently that changes in the residue sequences of TC molecules at the interface can affect the material mechanical strength considerably [13–15, 21]. Further, it has been shown earlier that moisture can play a major role in affecting the strength of such hybrid interfaces in biological materials [22–25].

Challenges lie in identifying nature's mechanisms behind imparting such properties and its pathways in fabricating and optimizing these composites. The route frequently acquired by nature is embedding submicron or nanosized mineral particles in protein matrix in a well-organized hierarchical arrangement. The key here is the formation of large amount of precisely and carefully designed organic–inorganic interfaces and synergy of mechanisms acting over multiple scales to distribute loads and damage, dissipate energy, and resist change in properties owing to damages such as cracking.

The length scale and complexity of microstructure of hybrid interfaces in biological materials make it difficult to study them and understand the underlying mechanical principles which are responsible for their extraordinary mechanical performance. For this reason, the governing mechanisms for the mechanical behavior for such biomaterials are not understood completely. At the same time, such building block level understanding is not only important for the evolution of biological materials science but vital to the development of bioinspired materials. Attractive mechanical properties of hard biological materials have given birth to biomimetism and new biomaterials, which have been reviewed in past by many [4, 18, 27–34].

This chapter presents a brief overview of the role of interfacial structural design and interfacial forces in imparting superior mechanical performance to hard biological materials. Focus is on understanding the underlying engineering principles of nature's materials for use in biomedical engineering and biomaterial development.

5.2   Hard Biological Materials

Bone is an excellent example of stiff biological nanocomposite materials. From the mechanical strength viewpoint, its main constituents are TC molecules and biological hydroxyapatite mineral. Biological hydroxyapatite is a poorly crystalline impure form of the compound hydroxyapatite (chemical formula, $Ca_5(PO_4)_3OH$), containing constituents such as carbonate, citrate, magnesium, fluoride, and strontium [35, 36]. Despite the fact that TC is a soft phase and HAP is brittle, together they form a material of high mechanical strength and fracture toughness [37]. The structure of bone is organized over several length scales, with six to seven levels of hierarchy. Figure 5.1 shows the several hierarchical features from macroscopic to atomic scale. TC molecules assemble into collagen fibrils in a hydrated environment, which mineralize by formation of HAP crystals in the gap regions that exist due to the staggered arrangement. These mineralized collagen fibrils (MFCs) assemble together with extrafibrillar matrix to form next hierarchical layer of bone. While the structures at scales larger than MCFs vary for different bone types, MFCs are highly conserved, nanostructural primary building blocks of bone that are found universally [6, 38–40]. As shown in Fig. 5.1, at the fundamental scale, collagen fibrils are formed by staggered self-assembly of 300 nm long TC molecules, and mineral HAP occupies the gap regions along with water. Collagen fibril consists of some other constituents as well (such as non-collagenous proteins, glycans, mineral impurities, etc.). However, interfacial interactions between the TC and the HAP phases along with their hierarchical arrangement are thought to be the most important factors imparting high mechanical strength [41]. As shown using dotted and solid ellipses in Fig. 5.1, a staggered arrangement is responsible for possible tension–shear type of load transfer between TC molecules and HAP crystals in bone and in similar other biomaterials [42]. There have been growth- and mineralization-based explanations behind the existence of such an arrangement [10, 12, 43]. However, it may also be possible that the existence of such structural arrangement is driven by its important role in imparting high mechanical strength.

Nacre is another example of high toughness natural biomaterials found in the interior layers of the hard outer shells of mollusks. The exterior layers of the shell are typically brittle, but hard, and provide resistance to penetration from external impact [44, 45], while nacre, present in the inner layers of shell, provides toughening by dissipating the mechanical energy owing to its capability of undergoing large inelastic deformations [46]. The main constituents of nacre are aragonite (a crystallographic form of calcium carbonate, 95 % by volume) and elastic organic material (proteins such as chitin and polysaccharides). Nacre has a hierarchical brick-and-mortar structure (Fig. 5.2) with polygonal aragonite tablets as bricks and organic matrix as the mortar acting as the filler and adhesive between such tablets [47]. The aragonite tiles are roughly 0.5 µm thick and 5–8 µm in diameter. In some of the cases, like red abalone, nacre shows some level of organization across different layers, where tablets are stacked in columns with some overlap between tablets from adjacent columns. The interface between the tablets is roughly 30 nm thick and is composed of organic materials [48], nanoasperities [46], and direct mineral bridges connecting two adjacent tablets [49, 50]. In fact, the tablets themselves are formed by aragonite nanograins (Fig. 5.2g), which again are separated by a very fine three-dimensional network of organic material [51, 52].

5.2.1 Role of Interfaces in Hard Biomaterial Mechanics

The high toughness in bone is attributed to the ability of its microstructure to dissipate deformation energy without propagation of crack. Different toughening mechanisms have been reported for bone [53], such as crack deflection and crack blunting at the interlamellar interfaces, formation of non-connected micro-cracks ahead of the crack tip, and crack bridging in the wake zone of the crack. Important contributions to high toughness and defect tolerance of natural biomineralized composites are believed to arise from the nanometer-scale structural motifs such as collagen fibrils in bone. When under loading, both mineral nanoparticles and mineralized fibrils deform at first elastically but to different degrees [40]. The hierarchical nature of bone deformation is exemplified by a staggered model of load transfer in bone matrix which is shown in Fig. 5.3. The long and thin (100–200 nm diameter) mineralized fibrils lie parallel to each other and are separated by a thin layer (1–2 nm thick) of extrafibrillar matrix [54]. When external tensile load is applied to the tissue, it is resolved into a tensile deformation of the mineralized fibrils and a shearing deformation in the extrafibrillar matrix. The mineralized phase provides strengthening and organic phase provides toughness. The deformation in mineralized matrix not only occurs at the collagen fibril level but also occurs at all other hierarchical levels and different length scales. The interaction of the mineralized and organic components produces a synergistic effect that enhances mechanical properties.

In nacre, aragonite tablets and organic material are organized in such a fashion that if a load is applied perpendicular to the plane of tablets, the organic material arrests the uncontrolled growth of crack by deforming inelastically and dissipating energy [55–57]. In this mechanism, the tablets approximately remain linear elastic, and the large deformations are provided by significant shearing of the organic–inorganic interface. The presence of organic material between nanograins in a single tablet might also provide some resistance to tablet sliding. However, this viscoelastic mechanism is active only in hydrated state of nacre. A comparison of measurement

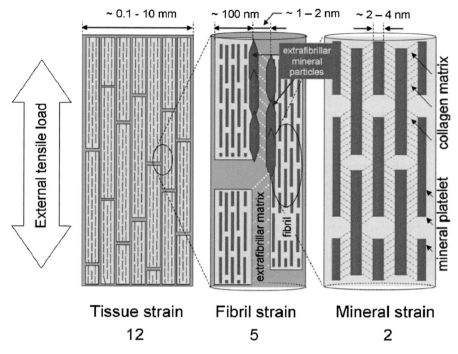

Fig. 5.3 Schematic model for bone deformation in response to external tensile load at three levels in the structural hierarchy: at the tissue level (*left*), fibril array level (*center*), and mineralized collagen fibrils (*right*). (*Center*) The stiff mineralized fibrils deform in tension and transfer the stress between adjacent fibrils by shearing in the thin layers of extrafibrillar matrix (white *dotted lines* show direction of shear in the extrafibrillar matrix). The fibrils are covered with extrafibrillar mineral particles, shown only over a selected part of the fibrils (*red hexagons*) so as not to obscure the internal structure of the mineralized fibril. (*Right*) Within each mineralized fibril, the stiff mineral platelets deform in tension and transfer the stress between adjacent platelets by shearing in the interparticle collagen matrix (*red dashed lines* indicate shearing qualitatively and do not imply homogeneous deformation) (adapted from [40])

of mechanical properties of nacre [3, 46, 55, 58–60] and mechanical properties of individual tablets of nacre using nanoindentation [51, 59, 61] has shown that both of them have properties similar to single-crystal aragonite, which is stiff but brittle. Considering that 95 % of nacre is aragonite, dry nacre behaves very much similar to pure aragonite because organic material becomes brittle. Apart from the above mechanism, few other mechanisms and factors are also proposed, which contribute to hardening, energy dissipation, and damage distribution. Similar to protein unfolding domains in bone [2, 62], it is suggested that proteins in nacre also have folded modules and act as high-performance adhesives, binding the tablets and possibly the nanograins together [56]. During deformation under loading, these biopolymers elongate in stepwise fashion and absorbing significant amount of energy. The interlocking between mineral platelets [63] increases toughness by progressively failing and limiting the catastrophic failure [64]. The waviness found in mineral tablets can be significant [4] and progressively locks tablet sliding, leading to a more difficult separation of tablets from their interfaces.

5.2.2 Modeling of TC–HAP and Generic Polymer–Ceramic Type Nanocomposites at Fundamental Length Scales

Several studies have been performed to understand the mechanical behavior of bone and bone-like nanocomposite materials. Deformation, damage, and failure mechanisms at nanoscale seem to be as important as the mechanisms at macroscopic length scale. The mechanical behavior of bone with a view to understand the role of TC molecules and HAP mineral has been earlier analyzed using experiments, modeling, and simulations. Experimental approaches have focused on analyzing tensile failure of single collagen fibers and fibrils [54, 65, 66] and on analyzing structural features at the nanoscale and its relation with the bone tissue failure [2, 67, 68]. Eppell and coworkers used a microelectromechanical device to obtain the first stress–strain curve of an isolated collagen fibril. They reported a low-strain Young's modulus of 0.5 GPa and a high-strain Young's modulus of ~12 GPa for the collagen fibril. Hansma and coworkers found that collagen in bone contains sacrificial bonds which may be partially responsible for the toughness of bone. The time needed for these sacrificial bonds to reform after pulling correlated with the time needed for bone to recover its toughness as measured by the atomic force microscope-based indentation testing. Modeling using the continuum approaches has focused on understanding the role played by the shear strength of TC molecules and the tensile strength of HAP mineral in flaw-tolerant hierarchical structural design of biomaterials [38, 69].

Explicit atomistic simulations using methods such as MD allow us to work at the lowest fundamental length scale and reveal structure–property relationships as well as mechanisms at the building block level. Previously such MD schemes have focused on understanding the mechanical behavior and properties of TC molecules in different structural configurations [70–72], on understanding the hierarchical organization of TC molecules into collagen fibrils and its effect on mechanical properties [73, 74], on understanding the properties of hydrated TC molecules [24, 75], and on understanding TC molecule stability with respect to changes in residue sequences [76].

5.2.2.1 Analytical Modeling

One of the interesting studies performed by [37] proposes an analytical model called tension–shear chain model (TSC), explaining the load transfer mechanism in generic hierarchical nanocomposite material systems such as bone and bone-like materials [69]. Under uniaxial tension, the path of load transfer in the staggered nanostructure follows a tension–shear chain with mineral platelets under tension

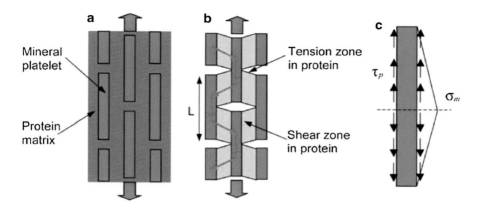

Fig. 5.4 Models of biocomposites. (**a**) Perfectly staggered mineral inclusions embedded in protein matrix. (**b**) A tension–shear chain model of biocomposites in which the tensile regions of protein are eliminated to emphasize the load transfer within the composite structure. (**c**) The free body diagram of a mineral crystal. Images are borrowed from [37]

and the soft matrix under shear (Fig. 5.4). The load transfer is largely accomplished by the high shear zones of protein between the long sides of mineral platelets. Under an applied tensile stress, the mineral platelets carry most of the tensile load while the protein matrix transfers the load between mineral crystals via. The mineral crystals have large aspect ratios and are much harder than the soft protein matrix, and the tensile zone in protein matrix near the ends of mineral crystals is assumed to carry no mechanical load. The work has addressed the prevalence and importance of nanoscale features in biological composites and also developed methodology for estimating interfacial shear strength, fracture localization width, and characteristic length of mineral crystal. A fractal-based analysis of TSC model is also performed to gain understanding for the need of different hierarchical structural levels. The analysis shows that nanometer size of brittle mineral crystals is essential to make them insensitive to crack-like flaws [7].

The observation that the building block level structures in biological materials are always at the nanoscale has intrigued researchers for a long time. Gao and coworkers [7] found that the nanoscale dimension of a mineral may be the result of fracture strength optimization. The fracture strength becomes sensitive to crack-like flaws when the mineral size exceeds nanoscopic length scale and fails by flaw propagation under stress concentration at crack tips. Perhaps nature finds this secret of optimum fracture strength and maximum flaw tolerance by evolution and hides mineral defects by designing the fundamental biomaterial structure at the nanoscale to achieve the robustness needed for survival.

Among collagenous tissues such as bone and tendon, the common structural feature is the mineralized collagen fibril acting as a building block. Accurate modeling of such mineralized collagen fibril is important and will require the right measurements of the dimensions, mechanical properties, and relative arrangement of the nano-length-scale constituents. Collagen molecules form ~100–200 nm diameter fibrils, with mineral particles inside and outside the surface [77]. Most studies describe the mineral particles to be plate shaped with a wider range of geometrical dimensions. The thickness of the platelets can range from 2 to 7 nm, the length from 15 to 200 nm, and the width from 10 to 80 nm [39, 40]. Tropocollagen molecule of course is one of the fundamental building blocks and has a length of ~300 nm. Both HAP mineral and tropocollagen molecules are arranged in a staggered fashion (as shown in Fig. 5.1f) with a gap of ~67 nm between two successive TC molecules creating space for mineral nucleation and growth [6]. With the current state-of-the-art methods, complete measurements for anisotropic stress–strain–strength properties for biological hydroxyapatitic mineralite and TC molecule are not available. Without these measurements, current models generally utilize elastic moduli measured for hydroxyapatitic polycrystallites. This does not lead to an accurate modeling of mineralized fibril because the hydroxyapatitic mineral in calcified tissue is usually nanocrystals.

5.2.2.2   Atomistic Modeling

In the structural studies of hard biological materials, it is observed that the mineral crystals tend to have a preferential alignment with respect to the loading direction and longitudinal axis of the polypeptide molecules [10–12, 78]. For example, in bone, both HAP crystal largest dimension (c-axis) and TC longitudinal axis are aligned along the maximum load-bearing direction [79] with a specific staggered structural arrangement (Figs. 5.1 and 5.5), permitting a maximum contact area. Resulting nanoscale interfacial interactions between TC phase and HAP phase is a strong determinant of the strength of such materials. Recent works by [9, 19, 20, 80, 81] present a mechanistic understanding of such interfacial interactions by examining idealized TC and HAP interfacial biomaterials. A three-dimensional atomistic modeling framework is developed which combines both organic and inorganic cells together to form a supercell as shown in Fig. 5.5. Two levels of hierarchical simulation supercells corresponding to level n and level $n+1$ (Fig. 5.5) were generated with two different possible types of interfaces (Fig. 5.1f). Secondary supercells are formed by embedding corresponding primary supercells in TC matrix. Two different loadings, transverse and longitudinal, were applied to all supercells. For this purpose, quasi-static mechanical deformation of each supercell is performed and characteristic stress–strain curves were obtained (Fig. 5.6) [9, 19]. Also, both tensile and compressive loading cases were considered and compared (Fig. 5.6) [20]. Furthermore, effect of hydration on such TC–HAP biomaterials was also studied. For failure analysis, TSC model [37, 42] was used to estimate the interfacial shear strength and fracture localization zone width.

Young's modulus value for TC molecule was obtained to be ~9 GPa, which is a fairly good estimate and lies in the range of modulus values obtained by other researchers [9]. The analyses confirm that the relative alignment of TC molecules with respect to the HAP mineral surface such that the interfacial contact area is maximized and optimal direction of applied loading with respect to the TC–HAP interface orientation

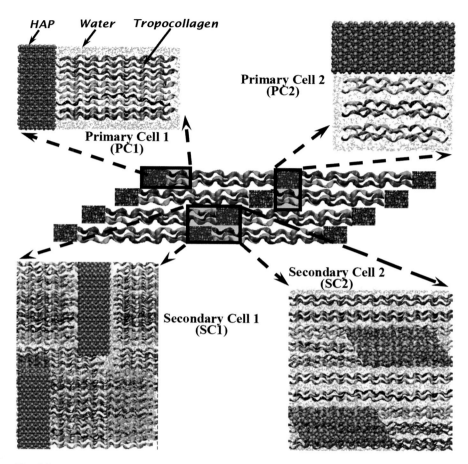

Fig. 5.5 A schematic showing the derivation of PC1, PC2, SC1, and SC2 cells from the staggered and layered assembly. In PC1 and SC1 cells, tropocollagen molecules are aligned in a direction parallel to the *c*-axis of hydroxyapatite crystals. In PC2 and SC2 cells, tropocollagen molecules are aligned in a direction normal to the longitudinal *c*-axis of hydroxyapatite supercell. Dimensions of hydroxyapatite crystals are approximately the same in all cells. Tropocollagen molecules are all shown in multicolor segments and water molecules are shown in *cyan*

are important factors that contribute to making nanoscale staggered arrangement a preferred structural configuration in such biological materials. The analyses also point out that such an arrangement results in higher interfacial strength as well as higher fracture strength. In addition, such TC–HAP nanocomposite shows toughening and strain-hardening behavior, which is attributed to the reconstitution of columbic interactions between TC and HAP at the interface during sliding. The dominant tensile failure mechanism at the HAP–TC interface is simply the interfacial separation of TC and HAP without significant initial HAP deformation (Fig. 5.6). The NH_3^+ and COO^- groups in TC molecules are strongly attracted to the ions in HAP surface (Ca^{2+}, PO_4^{3-}, and OH^- ions) [82]. Since TC is a flexible chain-like molecule, it elongates on applied

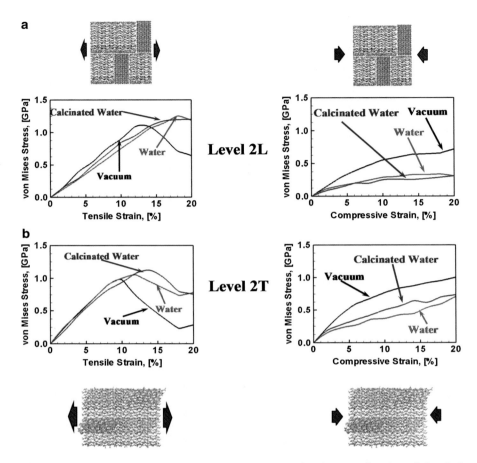

Fig. 5.6 Tensile and compressive von Mises stress vs. strain plots as a function of chemical environment in the case of (**a**) supercell layer 2L loaded in longitudinal direction and (**b**) supercell level 2T loaded in transverse direction

deformation but cleaves off after the point when it is fully stretched. Such cleavage results in local nanoscale interfacial failure.

Most biological materials in physiological systems contain water as one of their constituents. It has been observed that the presence of water molecules in the TC–HAP biocomposites enhanced overall composite mechanical strength (Fig. 5.6a) [9]. This is attributed to water molecule's affinity for charged surfaces, such as HAP surface [82], and NH_3^+ and COO^- groups in TC, owing to its polar nature and capability of make strong hydrogen bonds. As a result, water acts as an electrostatic bridge between HAP surface and TC molecules and strengthens the TC–HAP interface. This strengthening especially plays an important role whenever there is a relative sliding that occurs between HAP surface and TC molecules at the interface. Previous studies have shown that hydration has a stabilizing effect on the collagen triple helix [22], and solvated TC molecule requires more energy to untie from the HAP surface [83]. Similar interaction behavior of water at protein–mineral interface is found in nacre as well [84, 85]. It also acts as glue between TC–TC interactions [24] and, thereby, delays the failure of the overall system. Another interesting finding that emerged out of a comparison between stress–strain curves for two different hierarchical levels is that the failure of such biocomposites is predominantly strain dependent and not a function of ultimate strength.

Bone diseases such as osteogenesis imperfecta are marked by extreme bone fragility and are associated with point mutations in the tropocollagen molecule. Also, there has been a debate as to whether the HAP crystals in bone tissues are plate shaped or needle shaped. Hence, a further investigation into the effect of change in mineral crystal shape and effect of change in TC residue sequences on the mechanical strength of TC–HAP biomaterials was performed. Results show that TC–HAP interface shear strength increases as the side group complexity and heterogeneity of residues increase in the TC–HAP nanocomposite, and the plate-shaped crystals are overall better in resisting load as compared to needle-shaped HAP crystal case [86]. However, the effect of change in mineral crystal morphology has a stronger effect on the mechanical strength of the TC–HAP biocomposites, as compared to change in TC residue sequences [81]. This suggests that probably mutations in TC manifest its effect by changing the mineral crystal morphology and distribution during nucleation and growth period over the lifetime of the animal.

5.3 Bioengineering and Biomimetics

Materials science researchers are intensively seeking to learn the engineering and design principles from nature's hard biological materials to develop novel high-performance man-made composites. Many experimental have been carried out in this direction. Some of these efforts are discussed here. The ordered brick-and-mortar arrangement is considered to be the key strength and toughness-determining structural feature of nacre. It has also been shown that the ionic cross-linking of tightly folded macromolecules is equally important. Tang and coworkers [87] have demonstrated that both structural features can be reproduced by sequential deposition of polyelectrolytes and clays (Fig. 5.7). The resulting organic–inorganic hybrid reproduced both the brick-and-mortar arrangement and the crucial effect of the sacrificial bonds. The ultimate tensile strength and Young's modulus of such a composite approached that of nacre and lamellar bones, respectively. This makes such materials potential candidates for bone implant applications. Further, this layer-by-layer assembly is used to develop the antibacterial biocompatible version of the above organic–inorganic material [88].

The architecture of nacre is such that if a stress normal to the platelet plane is applied, then the ductile organic fraction that glues the crystals will prevent an uncontrolled crack growth [57]. This viscoelastic dissipative process takes place mainly in the hydrated state, and dry nacre behaves very much similar to pure aragonite

Fig. 5.7 Examples of "artificial" nacre, clay/polyelectrolyte-layered material (image adapted from [87])

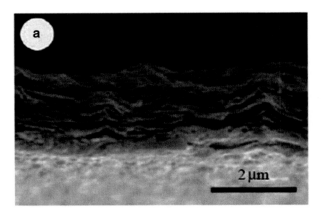

because the organic material becomes brittle [4]. The transition from a brittle (glassy) state to a more elastomeric behavior was observed in a polysaccharide (chitosan) upon hydration using nonconventional dynamic mechanical analysis, in which a glass transition could be detected in intermediate hydration levels [89]. Photoacoustic Fourier transform infrared spectroscopy results suggest that the water present at the nanograin interfaces also contributes significantly to the viscoelasticity of nacre [90]. It was also suggested that proteins composing the organic matrix, such as Lustrin A, exhibit a highly modular structure characterized by a multidomain architecture with folded modules and act as a high-performance adhesive, binding the platelets together [91]. Upon increasing stress, these biopolymers elongate in a stepwise manner as folded domains or loops are pulled open, with significant energy requirements to unfold each individual module — it was claimed that this modular elongation mechanism may contribute to the amazing toughness properties in nacre [91]. Other toughening pathways are also present such as the presence of asperities onto the surface of the aragonite tablets [92], mineral bridges [49], interlocks [93], waved surfaces [4], nanoscale nature and organization of the building blocks constituting nacre [94], and rotation and deformation of aragonite nanograins [95].

The overall function and properties of hard biological composite materials are a complicated coupled function of the structural arrangement at different hierarchical levels and individual constituent material properties. In general, it is impractical to separate structure and properties of such biocomposites to explain the material behavior. The success in mimicking even relatively simple structures, such as nacre, has been relatively modest yet. The best examples [87, 88] are far away from providing a millimeter-thick artificial nacre layer with greatly enhanced fracture toughness, a relatively simple task a mollusk has been routinely doing during its life cycle. Unlike artificial nanocomposite materials built using optimal interfacial arrangements, biological materials show remarkable strength compared to their relatively weaker constituents. In fact, sometimes artificial nanocomposites can turn out to be weaker than its constituents. Two important aspects of the biological material design arise here, which differentiate the artificial and natural nanocomposites:

(1) prevalence of carefully designed organic–inorganic interfaces at almost all hierarchical levels and (2) multilevel hierarchy with different microstructure designs at different levels. The first aspect points out that biological materials have extremely large amount of interfacial area. Moreover, these interfaces are formed by specific relative arrangement between the organic phases and inorganic phases for optimum load handling. Second, different microstructural designs at each level provide further optimization/customization parameters for desired mechanical performance. This suggests that an in-depth understanding of chemo-mechanics of such interfaces, their role in load handling, and failure mechanisms at the fundamental length scales is vital to the development of both the bioengineering and biomimetics fields.

5.4 Summary

Hard biological materials such as nacre and bone exhibit remarkable mechanical performance despite the fact that they are made up of relatively very weak constituents. The two main features identified for such behavior are interfacial nanostructural design and complex hierarchy. From an engineering standpoint, they are capable of inspiring a next generation of composite materials with high strength and toughness. This requires a clear understanding of nature's engineering design principles, fabrication pathways, and selection of appropriate materials for creating such biological composites. In terms of the underlying mechanical principles for structural design of these nanocomposites, quite a few have been suggested. For example, one principle is the alignment of mineral–protein interface along the loading directions. MD study of TC–HAP biomaterials shows that a composite is best poised to handle the load if the protein molecules are in contact with mineral crystals having their longitudinal axis parallel to the mineral surface and along the loading direction of the composite. The second principle is the staggered arrangement of hard mineral crystals in soft protein matrix, leading to a unique mechanism of load transfer where crystals bear the normal load and protein transfers the load via shear. The third principle is that the failure of such polymer–ceramic-type composites is dominantly peak strain dependent instead of peak strength. Another interesting observation is that such biomaterials become flaw tolerant at nanoscale due to special crack deflection and crack strengthening pathways. Also, the presence of moisture at the interface enhances the stability and strength of such biomaterials by supporting the cross-linking mechanism due to polar nature of water molecule.

One common feature which strongly stands out in most hard biological material structures is the presence of interfaces at multiple levels of hierarchy. It seems that nature has designed these interfaces for optimum multifunctional performance during the course of evolution. Interfacial forces play a key role during deformation and failure of such biomaterials. Such interfacial interaction between the soft phase and hard phase is responsible for redistribution of stresses and directly affects the toughness and strength of the biocomposite material. Further, the design of the polymer–mineral interface along with the critical length of mineral constituent also contributes potentially in strengthening the biocomposite against failure and in affecting the overall mechanical performance. Unlike nature which has a relatively restricted set of materials to choose from, engineers have more choices available for selecting materials to form composites. In order to capitalize on this advantage, a good understanding of nanoscale mechanics and structure–property function relationship of hard biological materials is vital. Therefore, characterization of such biological interfacial designs and interactions between the constituents are critical to the development of bioinspired materials, primarily because to swap materials in composite design, one should understand and predict the effects of specific material selections and design on the overall mechanical performance of biomaterials. This is becoming feasible with increasing experimental and modeling efforts in this active research area. However, further accurate measurements and analysis are required to capture the totality of all the factors for realistic biomimicking material development.

References

1. G. Niebur, http://www.nd.edu/~gniebur
2. G.E. Fantner et al., Sacrificial bonds and hidden length dissipate energy as mineralized fibrils separate during bone fracture. Nat. Mater. **4**, 612–616 (2005)
3. J.D. Currey, Mechanical-properties of mother of pearl in tension. Proc. R. Soc. Lond. B Biol. Sci. **196**, 443–463 (1977)
4. F. Barthelat, Biomimetics for next generation materials. Philos. Trans. R. Soc. A Math. Phys. Eng. Sci. **365**, 2907–2919 (2007). doi:10.1098/rsta.2007.0006
5. J.Y. Rho, L. Kuhn-Spearing, P. Zioupos, Mechanical properties and the hierarchical structure of bone. Med. Eng. Phys. **20**, 92–102 (1998)
6. S. Weiner, H.D. Wagner, The material bone: structure mechanical function relations. Annu. Rev. Mater. Sci. **28**, 271–298 (1998)
7. H.J. Gao, B.H. Ji, I.L. Jager, E. Arzt, P. Fratzl, Materials become insensitive to flaws at nanoscale: lessons from nature. Proc. Natl. Acad. Sci. U. S. A. **100**, 5597–5600 (2003). doi:10.1073/pnas.0631609100
8. R. Vaia, Polymer nanocomposites: status and opportunities. MRS Bull. (USA) **26**, 394–401 (2001)
9. D.K. Dubey, V. Tomar, Role of the nanoscale interfacial arrangement in mechanical strength of tropocollagen-hydroxyapatite based hard biomaterials. Acta Biomater. **5**, 2704–2716 (2009). doi:10.1016/j.actbio.2009.02.035
10. W.J. Landis, K.J. Hodgens, J. Arena, M.J. Song, B.F. McEwen, Structural relations between collagen and mineral in bone as determined by high voltage electron microscopic tomography. Microsc. Res. Tech. **33**, 192–202 (1996)
11. W.J. Landis et al., Mineralization of collagen may occur on fibril surfaces: evidence from conventional and high-voltage electron microscopy and three-dimensional imaging. J. Struct. Biol. **117**, 24–35 (1996)
12. P. Fratzl, N. Fratzlzelman, K. Klaushofer, G. Vogl, K. Koller, Nucleation and growth of mineral crystals in bone studied by small-angle X-ray scattering. Calcif. Tissue Int. **48**, 407–413 (1991)
13. D.K. Dubey, V. Tomar, Role of the nanoscale interfacial arrangement in mechanical strength of tropocollagen-hydroxyapatite based hard biomaterials. Acta Biomater. **5**, 2704–2716 (2009). doi:10.1016/j.actbio.2009.02.035
14. D.K. Dubey, V. Tomar, Understanding the influence of structural hierarchy and its coupling with chemical environment on the strength of idealized tropocollagen-hydroxyapatite biomaterials. J. Mech. Phys. Solid **57**, 1702–1717 (2009)

15. D.K. Dubey, V. Tomar, Effect of tensile and compressive loading on hierarchical strength of idealized tropocollagen-hydroxyapatite biomaterials as a function of chemical environment. J. Phys. Condens. Matter **21**, 205103 (2009)

16. P. Fratzl, R. Weinkamer, Nature's hierarchical materials. Prog. Mater. Sci. **52**, 1263–1334 (2007)

17. M.A. Meyers, P.Y. Chen, A.Y.M. Lin, Y. Seki, Biological materials: structure and mechanical properties. Prog. Mater. Sci. **53**, 1–206 (2008). doi:10.1016/j.pmatsci.2007.05.002

18. M.E. Launey, R.O. Ritchie, On the fracture toughness of advanced materials. Adv. Mater. **21**, 2103–2110 (2009). doi:10.1002/adma.200803322

19. D.K. Dubey, V. Tomar, Understanding the influence of structural hierarchy and its coupling with chemical environment on the strength of idealized tropocollagen–hydroxyapatite biomaterials. J. Mech. Phys. Solid **57**, 1702–1717 (2009). doi:10.1016/j.jmps.2009.07.002

20. D.K. Dubey, V. Tomar, The effect of tensile and compressive loading on the hierarchical strength of idealized tropocollagen-hydroxyapatite biomaterials as a function of the chemical environment. J. Phys. Condens. Matter **21**, 205103 (2009). doi:10.1088/0953-8984/21/20/205103

21. M.J. Buehler, Nanomechanics of collagen fibrils under varying cross-link densities: atomistic and continuum studies. J. Mech. Behav. Biomed. Mater. **1**, 59–67 (2008)

22. A.D. Simone, L. Vitaglaino, R. Berisio, Role of hydration in collagen triple helix stabilization. Biochem. Biophys. Res. Commun. **372**, 121–125 (2008)

23. R. Bhowmik, K.S. Katti, D.R. Katti, Influence of mineral-polymer interactions on molecular mechanics of polymer in composite bone biomaterials. Mater. Res. Soc. Symp. Proc. **978**, 6 (2007). Paper #: 0978-GG09I4-0905-FF0909-0905

24. D. Zhang, U. Chippada, K. Jordan, Effect of the structural water on the mechanical properties of collagen-like microfibrils: a molecular dynamics study. Ann. Biomed. Eng. **35**, 1216–1230 (2007)

25. N.M. Neves, J.F. Mano, Structure/mechanical behavior relationships in crossed-lamellar sea shells. Mater. Sci. Eng. C Biomim. Supramol. Syst. **25**, 113–118 (2005). doi:10.1016/j.msec.2005.01.004

26. G.M. Luz, J.F. Mano, Biomimetic design of materials and biomaterials inspired by the structure of nacre. Philos. Trans. R. Soc. A Math. Phys. Eng. Sci. **367**, 1587–1605 (2009). doi:10.1098/rsta.2009.0007

27. M. Sarikaya, C. Tamerler, A.K.Y. Jen, K. Schulten, F. Baneyx, Molecular biomimetics: nanotechnology through biology. Nat. Mater. **2**, 577–585 (2003)

28. B.D. Ratner, S.J. Bryant, Biomaterials: where we have been and where we are going. Annu. Rev. Biomed. Eng. **6**, 41–75 (2004). doi:10.1146/annurev.bioeng.6.040803.140027

29. C. Sanchez, H. Arribart, M.M.G. Guille, Biomimetism and bioinspiration as tools for the design of innovative materials and systems. Nat. Mater. **4**, 277–288 (2005). doi:10.1038/nmat1339

30. P. Fratzl, Biomimetic materials research: what can we really learn from nature's structural materials? J. R. Soc. Interface **4**, 637–642 (2007). doi:10.1098/rsif.2007.0218

31. B. Bhushan, Biomimetics: lessons from nature—an overview. Philos. Trans. R. Soc. A Math. Phys. Eng. Sci. **367**, 1445–1486 (2009). doi:10.1098/rsta.2009.0011

32. M.E. Launey et al., Designing highly toughened hybrid composites through nature-inspired hierarchical complexity. Acta Mater. **57**, 2919–2932 (2009). doi:10.1016/j.actamat.2009.03.003

33. E. Munch et al., Tough, bio-inspired hybrid materials. Science **322**, 1516–1520 (2008). doi:10.1126/science.1164865

34. L.C. Palmer, C.J. Newcomb, S.R. Kaltz, E.D. Spoerke, S.I. Stupp, Biomimetic systems for hydroxyapatite mineralization inspired by bone and enamel. Chem. Rev. **108**, 4754–4783 (2008). doi:10.1021/cr8004422

35. T. Leventouri, Synthetic and biological hydroxyapatites: crystal structure questions. Biomaterials **27**, 3339–3342 (2006)

36. S.C. Cowin, *Bone Mechanics Handbook* (CRC Press, Boca Raton, FL, 2001)

37. B. Ji, H. Gao, Mechanical properties of nanostructure of biological materials. J. Mech. Phys. Solid **52**, 1963–2000 (2004)
38. I. Jager, P. Fratzl, Mineralized collagen fibrils: a mechanical model with a staggered arrangement of mineral particles. Biophys. J. **79**, 1737–1746 (2000)
39. P. Fratzl, H.S. Gupta, E.P. Paschalis, P. Roschger, Structure and mechanical quality of the collagen-mineral nano-composite in bone. J. Mater. Chem. **14**, 2115–2123 (2004)
40. H.S. Gupta et al., Cooperative deformation of mineral and collagen in bone at the nanoscale. Proc. Natl. Acad. Sci. U. S. A. **103**, 17741–17746 (2006). doi:10.1073/pnas.0604237103
41. H.S. Gupta et al., Nanoscale deformation mechanisms in bone. Nano Lett. **5**, 2108–2111 (2005)
42. B.H. Ji, A study of the interface strength between protein and mineral in biological materials. J. Biomech. **41**, 259–266 (2008)
43. H.R. Wenk, F. Heidelbach, Crystal alignment of carbonated apatite in bone and calcified tendon: results from quantitative texture analysis. Bone **24**, 361–369 (1999)
44. J.D. Currey, J.D. Taylor, The mechanical behaviour of some molluscan hard tissues. J. Zool. **173**, 395–406 (1974)
45. M. Sarikaya, I.A. Aksay, *Biomimetic, Design and Processing of Materials. Polymers and Complex Materials* (American Institute of Physics, Woodbury, NY, 1995)
46. R.Z. Wang, Z. Suo, A.G. Evans, N. Yao, I.A. Aksay, Deformation mechanisms in nacre. J. Mater. Res. **16**, 2485–2493 (2001)
47. M. Sarikaya, An introduction to biomimetics—a structural viewpoint. Microsc. Res. Tech. **27**, 360–375 (1994)
48. T.E. Schaffer et al., Does abalone nacre form by heteroepitaxial nucleation or by growth through mineral bridges? Chem. Mater. **9**, 1731–1740 (1997)
49. F. Song, X.H. Zhang, Y.L. Bai, Microstructure and characteristics in the organic matrix layers of nacre. J. Mater. Res. **17**, 1567–1570 (2002)
50. F. Song, X.H. Zhang, Y.L. Bai, Microstructure in a biointerface. J. Mater. Sci. Lett. **21**, 639–641 (2002)
51. X.D. Li, W.C. Chang, Y.J. Chao, R.Z. Wang, M. Chang, Nanoscale structural and mechanical characterization of a natural nanocomposite material: the shell of red abalone. Nano Lett. **4**, 613–617 (2004). doi:10.1021/nl049962k
52. M. Rousseau et al., Multiscale structure of sheet nacre. Biomaterials **26**, 6254–6262 (2005). doi:10.1016/j.biomaterials.2005.03.028
53. H. Peterlik, P. Roschger, K. Klaushofer, P. Fratzl, From brittle to ductile fracture of bone. Nat. Mater. **5**, 52–55 (2006)
54. H.S. Gupta et al., Nanoscale deformation mechanisms in bone. Nano Lett. **5**, 2108–2111 (2005)
55. A.P. Jackson, J.F.V. Vincent, R.M. Turner, The mechanical design of nacre. Proc. R. Soc. Lond. B Biol. Sci. **234**, 415–440 (1988)
56. B.L. Smith et al., Molecular mechanistic origin of the toughness of natural adhesives, fibres and composites. Nature **399**, 761–763 (1999)
57. T. Sumitomo, H. Kakisawa, Y. Owaki, Y. Kagawa, In situ transmission electron microscopy observation of reversible deformation in nacre organic matrix. J. Mater. Res. **23**, 1466–1471 (2008). doi:10.1557/jmr.2008.0184
58. F. Barthelat, H. Tang, P.D. Zavattieri, C.M. Li, H.D. Espinosa, On the mechanics of mother-of-pearl: a key feature in the material hierarchical structure. J. Mech. Phys. Solid **55**, 306–337 (2007). doi:10.1016/j.jmps.2006.07.007
59. F. Barthelat, C.M. Li, C. Comi, H.D. Espinosa, Mechanical properties of nacre constituents and their impact on mechanical performance. J. Mater. Res. **21**, 1977–1986 (2006). doi:10.1557/jmr.2006.0239
60. R. Menig, M.H. Meyers, M.A. Meyers, K.S. Vecchio, Quasi-static and dynamic mechanical response of Haliotis rufescens (abalone) shells. Acta Mater. **48**, 2383–2398 (2000)
61. B.J.F. Bruet et al., Nanoscale morphology and indentation of individual nacre tablets from the gastropod mollusc Trochus niloticus. J. Mater. Res. **20**, 2400–2419 (2005). doi:10.1557/jmr.2005.0273

62. J.B. Thompson et al., Bone indentation recovery time correlates with bond reforming time. Nature **414**, 773–776 (2001)

63. K.S. Katti, D.R. Katti, S.M. Pradhan, A. Bhosle, Platelet interlocks are the key to toughness and strength in nacre. J. Mater. Res. **20**, 1097–1100 (2005). doi:10.1557/jmr.2005.0171

64. K.S. Katti, D.R. Katti, Why is nacre so tough and strong? Mater. Sci. Eng. C Biomim. Supramol. Syst. **26**, 1317–1324 (2006). doi:10.1016/j.msec.2005.08.013

65. N. Sasaki, S. Odajima, Stress-strain curve and Young's modulus of a collagen molecule as determined by the X-ray diffraction technique. J. Biomech. **29**, 655–658 (1996)

66. S.J. Eppell, B.N. Smith, H. Kahn, R. Ballarini, Nano measurements with micro-devices: mechanical properties of hydrated collagen fibrils. J. R. Soc. Interface **3**, 117–121 (2005)

67. A.J. Hodge, J.A. Petruska, in *Aspects of Protein Structure. Proceedings of a Symposium*, ed. by G.N. Ramachandran (1963), pp. 289–300

68. P.J. Thurner et al., High-speed photography of compressed human trabecular bone correlates whitening to microscopic damage. Eng. Fract. Mech. **74**, 1928–1941 (2007)

69. H. Gao, Application of fracture mechanics concepts to hierarchical biomechanics of bone and bone-like materials. Int. J. Fract. **138**, 101–137 (2006)

70. M.J. Buehler, Atomistic and continuum modeling of mechanical properties of collagen: elasticity, fracture, and self-assembly. J. Mater. Res. **21**, 1947–1962 (2006)

71. M.J. Buehler, Molecular nanomechanics of nascent bone: fibrillar toughening by mineralization. Nanotechnology **18**, 295102–295110 (2007)

72. A.C. Lorenzo, E.R. Caffarena, Elastic properties, Young's modulus determination and structural stability of the tropocollagen molecule: a computational study by steered molecular dynamics. J. Biomech. **38**, 1527–1533 (2005)

73. M. Israelowitz, S.W.H. Rizvi, J. Kramer, H.P. von Schroeder, Computational modeling of type I collagen fibers to determine the extracellular matrix structure of connective tissues. Protein Eng. Des. Sel. **18**, 329–335 (2005)

74. M.J. Buehler, Nanomechanics of collagen fibrils under varying cross-link densities: atomistic and continuum studies. J. Mech. Behav. Biomed. Mater. **1**, 59–67 (2008)

75. J.W. Handgraaf, F. Zerbetto, Molecular dynamics study of onset of water gelation around the collagen triple helix. Proteins Struct. Funct. Bioinform. **64**, 711–718 (2006)

76. R.J. Radmer, T.E. Klein, Triple helical structure and stabilization of collagen-like molecules with 4(R)-hydroxyproline in the Xaa position. Biophys. J. **90**, 578–588 (2006)

77. T. Hassenkam et al., High-resolution AFM imaging of intact and fractured trabecular bone. Bone **35**, 4–10 (2004)

78. S. Weiner, Y. Talmon, W. Traub, Electron diffraction of mollusc shell organic matrices and their relationship to the mineral phase. Int. J. Biol. Macromol. **5**, 325–328 (1983)

79. W.J. Landis, M.J. Song, A. Leith, L. McEwen, B.F. McEwen, Mineral and organic matrix interaction in normally calcifying tendon visualized in 3 dimensions by high-voltage electron-microscopic tomography and graphic image-reconstruction. J. Struct. Biol. **110**, 39–54 (1993)

80. D.K. Dubey, V. Tomar, Role of hydroxyapatite crystal shape in nanoscale mechanical behavior of model tropocollagen-hydroxyapatite hard biomaterials. Mater. Sci. Eng. C Mater. Biol. Appl. **29**, 2133–2140 (2009). doi:10.1016/j.msec.2009.04.015

81. D.K. Dubey, V. Tomar, Effect of osteogenesis imperfecta mutations in tropocollagen molecule on strength of biomimetic tropocollagen-hydroxyapatite nanocomposites. Appl. Phys. Lett. **96**, 023701–023703 (2010)

82. A.S. Posner, R.A. Beebe, The surface chemistry of bone mineral and related calcium phosphates. Semin. Arthritis Rheum. **4**, 267–291 (1975)

83. R. Bhowmik, K.S. Katti, D.R. Katti, Influence of mineral-polymer interactions on molecular mechanics of polymer in composite bone biomaterials. Mater. Res. Soc. Symp. Proc. **978**, 6 (2007)

84. F. Barthelat, H.D. Espinosa, An experimental investigation of deformation and fracture of nacre-mother of pearl. Exp. Mech. **47**, 311–324 (2007). doi:10.1007/s11340-007-9040-1

85. P. Ghosh, D.R. Katti, K.S. Katti, Mineral proximity influences mechanical response of proteins in biological mineral-protein hybrid systems. Biomacromolecules **8**, 851–856 (2007)

86. D.K. Dubey, V. Tomar, Effect of changes in tropocollagen residue sequence and hydroxyapatite mineral texture on the strength of ideal nanoscale tropocollagen-hydroxyapatite biomaterials. J. Mater. Sci. Mater. Med. **21**, 161–171 (2010). doi:10.1007/s10856-009-3837-7
87. Z.Y. Tang, N.A. Kotov, S. Magonov, B. Ozturk, Nanostructured artificial nacre. Nat. Mater. **2**, 413–418 (2003). doi:10.1038/nmat906
88. P. Podsiadlo et al., Layer-by-layer assembly of nacre-like nanostructured composites with antimicrobial properties. Langmuir **21**, 11915–11921 (2005). doi:10.1021/la051284+
89. J. Benesch, J. Mano, R. Reis, Proteins and their peptide motifs in acellular apatite mineralization of scaffolds for tissue engineering. Tissue Eng. Part B Rev. **14**, 433–445 (2008)
90. D. Verma, K. Katti, D. Katti, Nature of water in nacre: a 2D Fourier transform infrared spectroscopic study. Spectrochim. Acta Part A Mol. Biomol. Spectrosc. **67**, 784–788 (2007)
91. B.A. Wustman, J.C. Weaver, D.E. Morse, J.S. Evans, Structure–function studies of the Lustrin a polyelectrolyte domains, RKSY and D4. Connect. Tissue Res. **44**(Suppl. 1), 10–15 (2003)
92. G.M. Luz, J.F. Mano, Biomimetic design of materials and biomaterials inspired by the structure of nacre. Philos. Trans. R. Soc. A **28**(367), 1587–1605 (2009)
93. D.R. Katti, P. Ghosh, S. Schmidt, K.S. Katti, Mechanical properties of the sodium montmorillonite interlayer intercalated with amino acids. Biomacromolecules **6**, 3267–3282 (2005)
94. P. Fratzl, R. Weinkamer, Nature's hierarchical materials. Prog. Mater. Sci. **52**, 1263–1334 (2007)
95. X. Li, Z.-H. Xu, R. Wang, In situ observation of nanograin rotation and deformation in nacre. Nano Lett. **6**, 2301–2304 (2006)

Chapter 6
Multiscaling for Molecular Models: Investigating Interface Thermomechanics

Abstract One of the most important aspects of understanding the influence of interfaces on natural material properties is the knowledge of how stress transfer occurs across the organic–inorganic interfaces. The multicomponent hierarchical structure of biomaterials results in organic–inorganic interfaces appearing at different length scales, i.e., between the basic components at the nanoscale, between the mineralized fibrils at the microscale, and between the layers of the multilayered structures at micro- or macroscale. For a given peak tensile strength of a given material, which position of total strength is attributed to interface strength? What is the contribution of interface sliding in time-dependent deformation observed in a simple tension test of a given material sample? This chapter focuses on addressing such questions using molecular simulations.

Keywords Interface effect • Effect of interface deformation • Interface properties • Interface creep • Mechanics of interface deformation

Biological materials have evolved over millions of years and are often found as complex composites with superior properties compared to their relatively weak original constituents. Such materials often combine two properties which are contradictory but essential for their functioning. A unique feature that determines their properties is interfacial interaction between organic and inorganic phases in the form of protein (e.g., chitin (CHI) or tropocollagen (TC))–mineral (e.g., calcite (CAL) or hydroxy-apatite (HAP)) interfaces.

In the structural studies of such biological materials, it is observed that at the meso-scale (~100 nm to few μm), the mineral crystals are preferentially aligned along the length of the organic phase polypeptide molecules in a hierarchical (e.g., staggered or Bouligand pattern) arrangement, Fig. 6.1 [1–5]. Interfaces are perceived to play a significant role in the stress transfer and the consequent improvements in stiffness and strength of such material systems. However, how exactly the change in interfacial chemical configuration implies change in mechanical properties in such materials is a big subject of debate. One of the most important aspects of understanding the influence of interfaces on natural material properties is the knowledge of how stress transfer occurs across the organic–inorganic interfaces. The multicomponent hierarchical structure of biomaterials results in organic–inorganic interfaces appearing at different length scales, i.e., between

© Springer Science+Business Media New York 2015
V. Tomar et al., *Multiscale Characterization of Biological Systems*,
DOI 10.1007/978-1-4939-3453-9_6

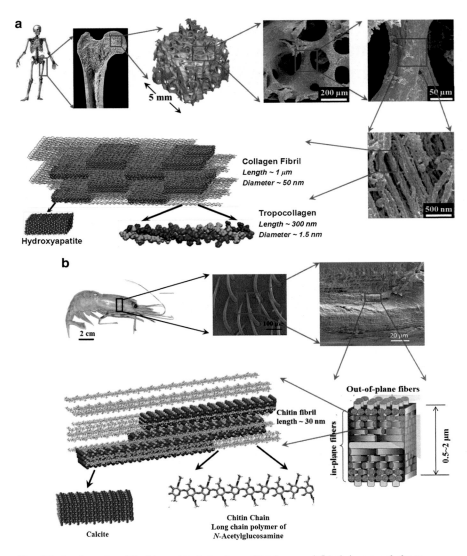

Fig. 6.1 A schematic of the hierarchical structure of (**a**) bone and (**b**) shrimp exoskeleton

the basic components at the nanoscale, between the mineralized fibrils at the microscale, and between the layers of the multilayered structures at micro- or macroscale. Earlier studies have focused on the role of interface-related mechanisms in determining overall mechanical deformation properties, the real aspect of stresses at interfaces while the mechanical deformation is going on, still remains unaddressed. In this context, important questions are as follows: For a given peak tensile strength of a given material, which position of total strength is attributed to interface strength? What is the contribution of

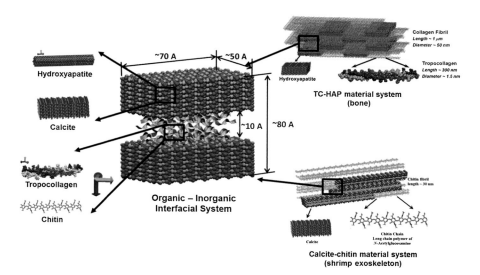

Fig. 6.2 A schematic showing configuration of the analyzed TC–HAP and CHI–CAL interfaces

interface sliding in time-dependent deformation observed in a simple tension test of a given material sample? Recently, simulation performed by Qu and Tomar [6] has pointed out some important aspects in answering such questions.

Figure 6.2 shows a schematic of the type of interfacial systems analyzed. CHI (chitin) and TC molecules are embedded in between CAL and HAP platelets, respectively. Such interfaces are then deformed under tensile and shear modes using established nonequilibrium molecular dynamics (NEMD) and steered molecular dynamics (SMD) schemes, with focus on measuring the interfacial shear strength in two separate deformation modes.

In order to analyze the effect of hydration on interface stress, water (WT) molecules are added to the interface region. The interface separation in Fig. 6.2 is chosen so as to have one layer of TC (or CHI) molecules in the interface region. Due to computational infeasibility of performing atomistic analysis of supercells with full-length TC (or CHI) molecules, only a segment of TC (or CHI) full-length molecule is used in the supercells. Stress–strain curve information is generated based on the well-known virial stress formulation using NEMD simulations. There are two loading directions (Fig. 6.3): direction along the molecule length (x-axis) and direction transverse to molecule length (y-axis). The simulated supercells are divided into slabs and three diagonal components of the pressure tensor in each slab are given in output (Fig. 6.3). The virial stress tensor of each slab and the overall system at the end of the equilibration is recorded as the stress tensor up to the point 20 % strain is achieved. The procedures make it possible to estimate how the measured stress of the loaded material system is distributed inside its interfacial regions and the stress between the interfaces can be obtained according to the behavior of the corresponding slab of the simulation system (Fig. 6.3).

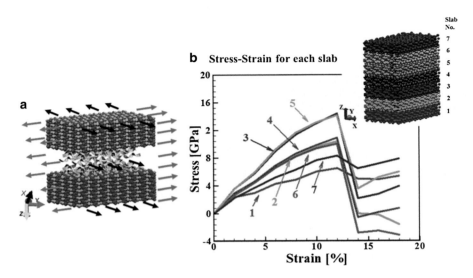

Fig. 6.3 (a) A schematic showing the loading condition of the interfacial material system and (b) stress–strain curve for each slab of the system from the *bottom* to the *top* layer. These plots are taken from Qu and Tomar [7]

NEMD simulations using the procedure are performed to predict interface strength by way of measuring stress–strain behavior of a thin block of atoms that are contained in the interfacial region. Such simulations cannot predict the effect of interface strength on interface separation mechanism. In order to understand such attribute, SMD simulations are performed. SMD simulations in the constant speed mode [8] were used to pull out the upper inorganic crystals (HAP or CAL) from the substrate inorganic crystals (HAP or CAL) in order to replicate the interfacial sliding process. Similar to the case of NEMD simulations, there are two loading directions: direction along the molecule length (*x*-axis) and direction transverse to molecule length (*y*-axis), Fig. 6.4. SMD force was applied to the center of mass of upper inorganic crystals in a chosen direction. The organic molecules (TC or CHI) and water molecules in the interface region were not under constraint. The substrate inorganic crystals were fixed on the bottom. In order to quantify the interface sliding process and failure in the interface region, a viscoplastic model [9] for interfacial sliding is introduced. The viscoplastic failure of the interfaces relates to the applied shear stress, τ, and to the shear velocity gradient (rate of shear deformation), $\frac{\partial V}{\partial d}$, after the yield stress, τ_0, is reached, as

$$\tau = \tau_0 + \mu \frac{\partial V}{\partial d}. \tag{6.1}$$

Here, μ is the shear viscosity of interfacial sliding, and d is defined in Fig. 6.4.

In order to study the effect of shear rate on the viscous behavior of the interfacial systems, differently large force increments were only applied to the HAP–TC–HAP

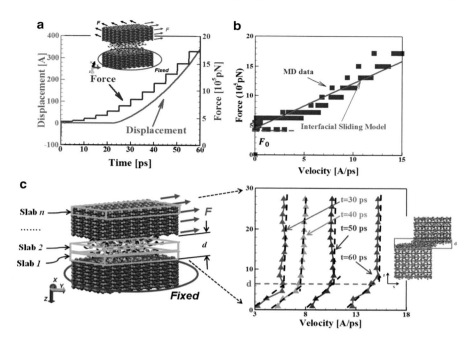

Fig. 6.4 (**a**) SMD force and displacement of the SMD structure as a function of time, with a schematic showing the loading condition of the SMD in constant force pulling mode, (**b**) curve fitting of the MD data with the viscoplastic model, and (**c**) velocity profile at different time steps. These plots are taken from Qu and Tomar [7]

(and CAL–CHI–CAL) system with one layer of TC (and CHI) molecules, due to the limited space to perform the interfacial sliding with different shear rate. The shear viscosity of the HAP–HAP interface (Fig. 6.4) is calculated as 0.0232 Pa s. The viscosity of the slurries of HAP was reported as ~0.01–1.6 Pa s earlier by experiments performed [10], and the viscosity of montmorillonite hydrate was reported as ~0.008 Pa s by MD simulations [11]. The thickness of the interface region, d, can be determined from the plot. The yield shear stress, τ_Y, is calculated from the critical force, F_0, obtained by the curve fitting and the interfacial area, A. The failure shear stress, τ_F, is calculated from the interface separation force divided by the interfacial area, A.

Shear stress plays an important role in the failure of the examined TC–HAP and CAL–CHI composite supercells. The lower and the upper bound of the shear strength of the interfacial material systems with organic interfaces or water interfaces could be defined by the yield shear stress, τ_Y, and the failure stress, τ_F, which are used to characterize the plastic shear deformation, where $\tau_Y < \tau_F$. Figure 6.5a displays the yield shear stress (τ_Y) calculated by means of SMD, the failure stress (τ_F) calculated using SMD, the interfacial shear strength of each of the TC–HAP interfacial material system shown in figure calculated using NEMD (σ_1), the shear strength of the organic TC phase region (or region of HAP cells with WT) (σ_3)

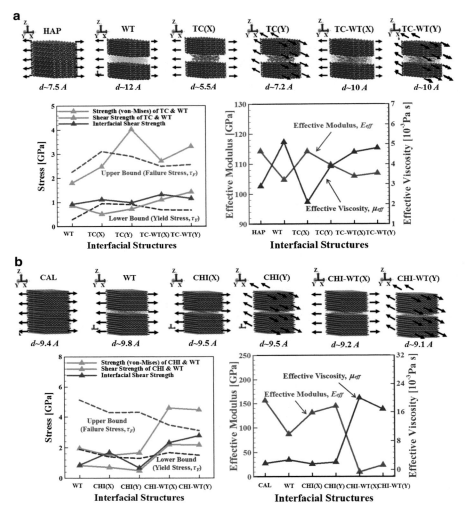

Fig. 6.5 Stress, Young's modulus, and effective viscosity, as functions of interfacial structures in the case of (**a**) TC–HAP interfaces and (**b**) CAL–CHI interfaces. These plots are taken from Qu and Tomar [7]

calculated using NEMD, and the tensile mechanical strength of the organic phase region (or region of HAP cells with WT) (σ_4) calculated using NEMD, as functions of the interfacial components.

Figure 6.5a also compares effective viscosity and Young's moduli in dependence of the interface type. The HAP supercell consists of two HAP cells placed over each other separated by layer distance corresponding to PO_4 using in planes (7.5 Å). The WT supercell consists of the same HAP supercells but now with water molecules separating the two. Figure 6.5b shows similar results in the case of CHI–CAL interfaces. In this case, the CAL supercell consists of two CAL cells placed over each other separated by layer distance corresponding to PO_4 using in planes (9.4 Å). The WT supercell consists of the same CAL supercells but now with water molecules separating the two.

Perhaps not surprisingly, the behavior of the two types of interfacial systems is quite similar to each other. The shear strength of organic phase (green line, σ_3) lies around the lower bounded line of the yield shear stress (blue dotted line, τ_Y). The organic phases are the main contributors of the interfacial shear strength based on Fig. 6.5, where the interfacial shear strength (gray line, σ_1) matches closely with, sometimes a little higher than, the shear strength of the organic phases (green line, σ_3). However, it is always below the upper bound defined by the failure stress (red dotted line, τ_F) which indicates the "catastrophic failure" of the interfaces. The mechanical strength of the organic phases (yellow line, σ_4) usually lies between the lower and upper bounded line because shear deformation is usually the main contributor of the mechanical behavior of organic phases. Those points which are beyond the upper bound concern hydrated organic interfaces (i.e., TC–WT or CHI–WT). This could be attributed to the much higher contribution of the shear interaction to the overall behavior as well as the higher shear viscosity of the interfacial material systems.

Whatever differing mechanisms are observable in the different investigated systems, all interfacial shear viscosities reported for the HAP–TC–HAP system are on the order of 10^{-2} Pa s. The viscosities obtained from our MD simulations are also much lower than experimental results of somehow related material systems, such as collagen gel with a viscosity of the order of 10^5 Pa s [12, 13]. The much higher shear rate in MD simulation leads to the measured lower viscosity values. Considered as the Newtonian fluid behavior, the shear rate ($\dot{\gamma}$) dependency of the viscosity (μ) of polymeric molecular structure is quite sensitive, i.e., increasing sharply as the shear rate decreases [14, 15]. Due to the computational capability of MD simulation, different shear rates varying from 10^7 to 10^9 1/s were performed on the HAP–TC–HAP system with one layer of TC molecules. Fitting with the widely used power law relation

$$\mu = B\dot{\gamma}^{n-1},\tag{6.2}$$

with the parameters $B = 42.49$ Pa s and $n = 0.7034$, however, it is still not enough to capture the full picture of the viscosity–shear relationship because shear rate is still much higher than that used in creep or stress relaxation experiments (i.e., 10^{-6} to 10^{-2} 1/s [14, 16] which is too low to be generated using MD simulation). The current study reports the viscous behavior of the biointerface systems at the infinite shear rate which can be used to estimate the properties beyond the observation range together with the material intrinsic property, i.e., zero-shear viscosity, μ_0, using the extrapolation methods, such as the cross model [14, 16]:

$$\mu = \mu_\infty + \frac{\mu_0 - \mu_\infty}{1 + \left(C\dot{\gamma}\right)^m}\tag{6.3}$$

Fig. 6.6 Plot showing viscosity as a function of shear rate; references: Knapp et al. [12]. Newman et al. [17]. Taken from Qu and Tomar [7]

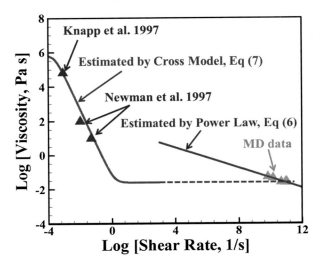

where μ_∞ is the infinite shear viscosity obtained from the MD simulation, C is the cross time constant, and m is known as the cross rate constant. With the viscosities of collagen materials at the lower shear rate from previous studies [12, 13, 17] and our infinity shear rate viscosity, the cross model parameters are obtained as $C = 3845.7$ s and $m = 2.002$. Figure 6.6 displays the viscosity versus shear rate behavior as a plot of $\log(\mu)$ versus $\log(\dot{\gamma})$. The cross model extrapolation captures the significant shear-thinning behavior of the material within the low shear rate region and gives the idea of upscaling the MD viscosity results with the decreasing of the shear rate. The overall shear rate-dependent viscous behavior of the material is predicted with the combination of the cross model and MD simulations.

Hard biological materials such as nacre, bone, and marine crustacean exoskeletons exhibit remarkable mechanical performance despite the fact that they are made up of relatively weaker constituents. In terms of the underlying mechanical principles for structural design of such materials, quite a few have been suggested. For example, one principle is the alignment of mineral–protein interfaces along the loading directions. MD study of TC–HAP biomaterials shows that a composite is best poised to handle the load if the protein molecules are in contact with mineral crystals having their longitudinal axis parallel to the mineral surface and along the loading direction of the composite. The second principle is the staggered arrangement of hard mineral crystals in soft protein matrix, leading to a unique mechanism of load transfer where crystals bear the normal load and protein transfers the load via shear. The third principle is that the failure of such polymer–ceramic type composites is dominantly peak strain dependent instead of peak strength. Also, the presence of moisture at the interface enhances the stability and strength of such biomaterials by supporting the cross-linking mechanism due to polar nature of water molecule.

One common feature which strongly stands out in most hard biological material structures is the presence of interfaces at multiple levels of hierarchy. It seems that nature has designed these interfaces for optimum multifunctional performance during the course of evolution. Interfacial forces play a key role during deformation and failure of such biomaterials. Inorganic phases in the material systems carry the uniaxial tensile loading, while the organic phases mainly carry the shear loading. Organic interfacial systems exhibit plastic shear deformation; the yield and failure shear stress define the lower and higher bound of the interfacial strength. Shear viscosity of the interfacial systems shows a highly shear rate-dependent behavior; however, the full picture of this behavior cannot be captured using MD simulation without the assistance from experimental technique and the extrapolation estimation. Interfacial interaction between the soft phase and hard phase is responsible for redistribution of stresses and directly affects the toughness and strength of the natural materials. Further, the design of the organic–inorganic interface along with the critical length of mineral constituent also contributes potentially in strengthening the biomaterials against failure and in affecting their overall mechanical performance.

A simple model combining Kelvin–Voigt and Maxwell model can be used to represent the viscous interfaces and calcite layers [7]. These models have been used by several researchers to model viscous behavior of bones and tissues [18, 19]. It has been reported [20, 21] that the viscous adhesive behavior of biointerfaces is dominated by sacrificial bonds between Ca^{2+} ions and organic fibers. This "glue" interface promotes energy dissipation during deformation of the material, and this mechanism works better when more positive ions, such as Ca^{2+}, are involved [20]. These viscous hydrated interfaces can act as the source of the macroscopic phenomenon of biomaterial viscoelasticity [22, 23]. Presence of the interface with higher viscosity is the main contributor of the toughening mechanisms to prevent catastrophic failure. At the nanoscale, the interfaces reduce the stiffness but improve the toughness of the biomaterial by affecting the stress distribution (more uniform along z-coordination), enhancing the shear contribution to the overall mechanical behavior and promoting energy dissipation required for viscoelastic deformation of the organic phases [24].

Figure 6.7 shows the representation of the Kelvin–Voigt and Maxwell model. It is assumed that calcite layers only contribute to the elastic part. The interface region contributes to both elastic and viscous parts of the system. Stress and strain

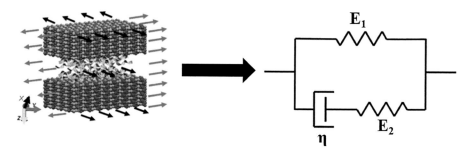

Fig. 6.7 Schematic of calcite and chitin interface based on the Kelvin–Voigt and Maxwell model. These plots are taken from Qu and Tomar [7]

Table 6.1 Parameters calculated by Kelvin–Voigt and Maxwell model and viscosity from MD simulations [7]

	Fit parameters from equation			MD simulation results	
	Calcite—modulus (GPa)	α-Chitin—modulus (GPa)	Viscosity (Pa s)	Modulus (GPa)	Viscosity (Pa s)
cal-chi-cal(x)	130	150	0.0036	130.90	0.0570
cal-chi-cal(y)	140	15	0.0011	144.70	0.0580
cal-chi-wt-cal(x)	12	0.15	0.0025	10.02	0.0460
cal-chi-wt-cal(y)	23	1.50	0.0004	23.86	0.0480

data for the analyzed models was calculated from the MD simulations. Modulus and viscosity for the MD simulation results are shown in Table 6.1.

Spring element with stiffness E_1 in Fig. 6.7 represents calcite blocks. Spring element E_2 is the stiffness of viscous interface and η represents viscosity of the interface. η is defined as the loss factor used to quantify damping performance. Higher η leads to more significant effect of the damping system on the deformation mechanism, more energy dissipation during the elastic deformation, more ductile behavior, and lower effective stiffness of the material. Solving this model, stress–strain relation can be expressed as [7]

$$\sigma = -\left(\frac{\eta}{E_2}\right)\frac{d\sigma}{dt} + E_1 * \varepsilon + \left(1 + \frac{E_1}{E_2}\right)\eta\frac{d\varepsilon}{dt}. \tag{6.4}$$

Here, σ is the stress, ε is strain, E_1 is stiffness of calcite, E_2 is stiffness of chitin, and η is viscosity of α-chitin interface. The stress–strain data from the analyzed models was fit into Eq. 6.4 to obtain values of parameters E_1, E_2, and η of Kelvin–Voigt and Maxwell system as given in Table 6.1. The values from both analytical model and the MD simulations show the same trend in viscosity. The difference of one order of magnitude in MD viscosity values is because of the model used for the calculation. MD model only considers shear stress contribution, while Eq. 6.4 considers the overall stress–strain behavior of the material.

The analyses of organic–inorganic biomaterials (e.g., bone, marine exoskeleton, nacre, etc.) based on the tension–shear chain models [25–27] with organic fibers (e.g., chitin) aligning in a direction parallel to the surface of inorganic crystals (e.g., calcite) pointed that the shear deformation between the soft matrix and inorganic crystals is the main stress transfer mechanism of such loaded biomaterials, and recent studies proposed that the layered water in crystal interfaces could be the source for biomaterial viscoelasticity [22, 23]. Therefore, the magnitude of viscosity reported in this study is a potentially important value to be used with analytical models to determine the biomaterial viscoelastic properties. The presence of water in interface decreases viscosity of the overall material as compared to interfaces without water. It also further affects the other elastic properties of material such as modulus and hardness.

References

1. W.J. Landis et al., Mineralization of collagen may occur on fibril surfaces: evidence from conventional and high-voltage electron microscopy and three-dimensional imaging. J. Struct. Biol. **117**, 24–35 (1996)
2. W.J. Landis, K.J. Hodgens, J. Arena, M.J. Song, B.F. McEwen, Structural relations between collagen and mineral in bone as determined by high voltage electron microscopic tomography. Microsc. Res. Tech. **33**, 192–202 (1996)
3. P. Fratzl, N. Fratzlzelman, K. Klaushofer, G. Vogl, K. Koller, Nucleation and growth of mineral crystals in bone studied by small-angle X-ray scattering. Calcif. Tissue Int. **48**, 407–413 (1991)
4. S. Weiner, Y. Talmon, W. Traub, Electron diffraction of mollusc shell organic matrices and their relationship to the mineral phase. Int. J. Biol. Macromol. **5**, 325–328 (1983)
5. A. Al-Sawalmih et al., Microtexture and chitin/calcite orientation relationship in the mineralized exoskeleton of the American lobster. Adv. Funct. Mater. **18**, 3307–3314 (2008)
6. T. Qu, V. Tomar, in *Proceedings of the Society of Engineering Science 51st Annual Technical Meeting, October 1–3, 2014*. ed. by A. Bajaj, P. Zavattieri, M. Koslowski, T. Siegmund (Purdue University Libraries Scholarly Publishing Services, West Lafayette, 2014)
7. T. Qu, V. Tomar, Influence of interfacial interactions on deformation mechanism and interface viscosity in chitin-calcite interfaces. Acta Biomater. **25**, 325–338 (2015). doi:10.1016/j.actbio.2015.06.034
8. J.C. Phillips et al., Scalable molecular dynamics with NAMD. J. Comput. Chem. **26**, 1781–1802 (2005)
9. S. Frankland, V. Harik, Analysis of carbon nanotube pull-out from a polymer matrix. Surf. Sci. **525**, L103–L108 (2003)
10. F. Lelievre, D. Bernache-Assollant, T. Chartier, Influence of powder characteristics on the rheological behaviour of hydroxyapatite slurries. J. Mater. Sci. Mater. Med. **7**, 489–494 (1996)
11. Y. Ichikawa, K. Kawamura, N. Fujii, T. Nattavut, Molecular dynamics and multiscale homogenization analysis of seepage/diffusion problem in bentonite clay. Int. J. Numer. Methods Eng. **54**, 1717–1749 (2002)
12. D.M. Knapp et al., Rheology of reconstituted type I collagen gel in confined compression. J. Rheol. **41**, 971–993 (1997)
13. V.H. Barocas, A.G. Moon, R.T. Tranquillo, The fibroblast-populated collagen microsphere assay of cell traction force—Part 2: Measurement of the cell traction parameter. J. Biomech. Eng. **117**, 161–170 (1995)
14. J.M. Dealy, J. Wang, *Melt rheology and Its Applications in the Plastics Industry* (Springer, Netherlands, 2013)
15. G. Bylund, T. Pak, *Dairy Processing Handbook* (Tetra Pak Processing Systems AB, Lund, 2003)
16. A. Franck, *Understanding Rheology of Thermoplastic Polymers* (TA Instruments, New Castle, DE, 2004)
17. S. Newman, M. Cloitre, C. Allain, G. Forgacs, D. Beysens, Viscosity and elasticity during collagen assembly in vitro: relevance to matrix-driven translocation. Biopolymers **41**, 337–347 (1997)
18. A. Gautieri, S. Vesentini, A. Redaelli, R. Ballarini, Modeling and measuring visco-elastic properties: from collagen molecules to collagen fibrils. Int. J. Nonlinear Mech. **56**, 25–33 (2013)
19. A. Gautieri, S. Vesentini, A. Redaelli, M.J. Buehler, Viscoelastic properties of model segments of collagen molecules. Matrix Biol. **31**, 141–149 (2012). doi:10.1016/j.matbio.2011.11.005
20. P.K. Hansma et al., Sacrificial bonds in the interfibrillar matrix of bone. J. Musculoskelet. Nueronal Interact. **5**, 313 (2005)
21. E. Thormann et al., Embedded proteins and sacrificial bonds provide the strong adhesive properties of gastroliths. Nanoscale **4**, 3910–3916 (2012)

22. L. Eberhardsteiner, C. Hellmich, S. Scheiner, Layered water in crystal interfaces as source for bone viscoelasticity: arguments from a multiscale approach. Comput. Methods Biomech. Biomed. Eng. **17**, 48–63 (2014)
23. M. Shahidi, B. Pichler, C. Hellmich, Viscous interfaces as source for material creep: a continuum micromechanics approach. Eur. J. Mech. A Solid **45**, 41–58 (2014)
24. M.A. Meyers, J. McKittrick, P.-Y. Chen, Structural biological materials: critical mechanics-materials connections. Science **339**, 773–779 (2013)
25. B. An, X. Zhao, D. Zhang, On the mechanical behavior of bio-inspired materials with non-self-similar hierarchy. J. Mech. Behav. Biomed. Mater. **34**, 8–17 (2014). doi:10.1016/j.jmbbm.2013.12.028
26. Z. Zhang, Y.-W. Zhang, H. Gao, On optimal hierarchy of load-bearing biological materials. Proc. R. Soc. B Biol. Sci. (2010). doi:10.1098/rspb.2010.1093
27. Z. Shuchun, W. Yueguang, Effective elastic modulus of bone-like hierarchical materials. Acta Mech. Solida Sin. **20**, 198–205 (2007)

Index

© Springer Science+Business Media New York 2015
V. Tomar et al., *Multiscale Characterization of Biological Systems*,
DOI 10.1007/978-1-4939-3453-9